Until recently, most of us thought of seaweeds as nothing but a nuisance, clinging to our legs as we swim in the ocean and stinking up the beach as they rot in the sun. With the ever-growing popularity of sushi restaurants across the globe, however, seaweeds are increasingly finding their way onto our plates. And even as we dine with delight on *maki*, *miso* soup, and seaweed salads, very few of us have any idea of the nutritional value of seaweeds. Here celebrated scientist Ole G. Mouritsen, drawing on his fascination with, and enthusiasm for, Japanese cuisine, champions seaweeds as a staple food while simultaneously explaining their biology, ecology, cultural history, and gastronomy.

Mouritsen takes readers on a comprehensive tour of seaweeds, describing what seaweeds actually are— marine algae, not plants—and how people of different cultures have utilized them since prehistoric times for a whole array of purposes—as food and fodder, for the production of salt, in medicine and cosmetics, as fertilizer, in construction, and for a number of industrial end uses, to name just a few. He describes the vast abundance of minerals, trace elements, proteins, vitamins, dietary fiber, and precious polyunsaturated fatty acids found in seaweeds, and provides instructions and recipes on how to prepare a variety of dishes that incorporate raw and processed seaweeds.

Approaching the subject from both a gastronomic and a scientific point of view, Mouritsen sets out to examine the past and present uses of this sustainable resource, keeping in mind how it could be exploited in the future. Because seaweeds can be cultivated in large quantities in the ocean in highly sustainable ways, they are ideal for battling hunger and obesity alike.

With hundreds of superb illustrations depicting the wealth of species, colors, and shapes of these marine algae, *Seaweeds: Edible, Available, and Sustainable* makes a strong case for granting these 'vegetables from the sea' a prominent place in our kitchens.

☙

Ole G. Mouritsen is a scientist and professor of biophysics with interests in research communication and the science of cooking.

Jonas Drotner Mouritsen is a graphic designer and owner of the design company Chromascope.

Mariela Johansen has Danish roots, lives in Canada, and holds an MA in humanities with a special interest in ancient Greece.

Seaweeds
edible, available
& sustainable

Ole G. Mouritsen

Seaweeds
edible, available
& sustainable

Photography, layout & design
Jonas Drotner Mouritsen

Translation & adaptation to English
Mariela Johansen

The University of Chicago Press
Chicago and London

OLE G. MOURITSEN is professor of biophysics with interests in research communication and the science of cooking.

Cover illustration: Jeff Foot, *Macrocystis pyrifera*, Catalina Island, California

The University of Chicago Press, Chicago 60637

The University of Chicago Press, Ltd., London

Based on a translation and modification for an international audience of the Danish edition *Tang. Grøntsager fra havet*, Nyt Nordisk Forlag Arnold Busck, Copenhagen, 2009, published with the support of Albani Fonden, Nordea Danmark-fonden, Konsul George Jorck og Hustru Emma Jorcks Fond, Møllerens Fond, Tuborgfondet, and Villum Kann Rasmussen Fonden.

www.seaweedbook.net

22 21   4 5

ISBN-13: 978-0-226-04436-1 (cloth)

ISBN-13: 978-0-226-04453-8 (e-book)

Library of Congress Cataloging-in-Publication Data

Mouritsen, Ole G., author.

[Tang. English]

Seaweeds : edible, available, and sustainable / Ole G. Mouritsen; photography, layout, and graphic design by Jonas Drotner Mouritsen; translation and adaptation by Mariela Johansen.

pages cm

Includes bibliographical references and index.

ISBN 978-0-226-04436-1 (cloth : alkaline paper) — ISBN 978-0-226-04453-8 (e-book) 1. Marine algae. 2. Marine algae as food. I. Mouritsen, Jonas Drotner. II. Johansen, Mariela, translator. III. Title.

QK570.2.M6813 2013

579.8'177—dc23

2012040240

Table of Contents

Technical and scientific details

Essays on seaweeds

Nihil vilior alga—there is nothing more worthless than seaweed.

(attributed to Virgil, 70–19 BCE)

Seaweed is a delicacy fit for the most honored guests, even for the king himself.

(Sze Teu, 600 BCE)

Preface

Although seaweeds are algae and not plants in the usual sense, humans have always eaten them. They grow in all of the Earth's coastal climatic zones, manifesting themselves in a wealth of species, colors, and shapes. In many countries in Asia, notably in China and Japan, seaweed products are an important dietary resource, which constitutes a substantial part of the total food intake. But at present they are a sadly overlooked source of nutrition in the Western world, where seaweeds are mainly used in the form of extracts as additives in foodstuffs and cosmetics. Most people living in Europe and the Americas are not even aware of their exposure to these products, to say nothing of the nutritional properties of marine algae.

My first encounter with seaweeds as edible items was as the skin-like outer layer of a sushi roll, one of the staples of Japanese cuisine. Until that time, seaweeds were for me, as for the majority of modern Westerners, simply useless things that washed up on the beach and stank horribly when they rotted in the sun, or that were a nuisance when one swam or dived in the ocean. As my interest in Japanese cuisine, which uses a large assortment of marine algae in soups, salads, seasoning mixtures, and sushi, literally went on to become a consuming passion, my fondness for seaweeds and seaweed products grew along with it. In the interval since I first discovered them as food, they have become something that my family and I eat just about every day in a whole slew of different ways. They are indispensable ingredients in our kitchen.

It was this fascination with seaweeds, as well as my desire to tell others about their uses as healthy and tasty food, that was initially the driving force that moved me to start writing. The book project evolved into a wish to share the other knowledge about them that I had acquired over the past two decades. My goal was to gather, in one place, a reader-friendly description of what seaweeds actually are and of how people of different cultures have utilized them since prehistoric times for a whole array of purposes—as food and fodder, for the production of salt, in medicine and cosmetics, as fertilizer, in construction, and for a number of industrial end use, to name just a few.

Approaching the subject from both a scientific and a gastronomic point of view, I set out to examine the past and present uses of this sustainable resource, keeping in mind how it could be exploited in the future. Focusing on the composition and consumption of seaweeds, the book makes the case for their exceptional nutritional value and includes recipes and instructions on how to prepare, in the home kitchen, a variety of dishes that incorporate raw and processed seaweeds. In addition, the book presents some examples of the ways in which innovative chefs have introduced seaweeds into cutting-edge gastronomy and elevated them to the level of fine dining.

Seaweeds can be used directly—either raw, dried, or cooked. While they have relatively few calories, seaweeds contain a vast abundance of important minerals, trace elements, proteins, and vitamins, as well as healthy dietary fiber and vital oils and fats. At this time, a couple of hundred different species of algae are consumed by people all over the world; in this volume we will have a closer look at about twenty of these. As they can both be harvested in the wild and cultivated in large quantities in the ocean in a sustainable manner, there is little doubt that seaweeds will, in the future, come to constitute a much greater proportion of the global food intake.

The more I progressed with the narrative in this book, the more I became intrigued with the notion that food from the sea was the most important element in the diet of Stone Age man. Seaweeds, like fish and shellfish, are 'brain food' and have played a vital role in the evolution of the human nervous system. It is possible that many different diet-related lifestyle diseases—known by the umbrella term 'metabolic syndrome' and including cardiovascular problems, cancer, obesity, diabetes, and a number of psychological illnesses—are due to the fact that genetically we are not well-suited for the typical calorie-rich Western diet, which features large quantities of white carbohydrates, a poor balance of vitamins and minerals, little dietary fiber, and many saturated fats. Reintroducing these 'vegetables from the sea', which were eaten by our ancestors, into our normal diets could bring about a welcome change in our Western eating patterns and might help to decrease the prevalence of metabolic syndrome in the general population.

Eating more marine algae could also be part of the solution to the problem of finding additional sustainable ways of providing food for a hungry planet. The oceans are, in a sense, the last places on Earth where humans can still push the boundaries. But this requires research and development of new methods and technologies for the cultivation, harvesting, and processing of the seaweed varieties with which we are already familiar, as well as with some entirely new ones. Most of all, it will also necessitate insightful planning that respects the ecological cycles of the oceans and the Earth and takes into account the needs of the populations that live along the coastlines and depend on the resources of the sea.

An evolution toward a better and more future-oriented exploitation of marine resources rests on a number of vital factors: disseminating facts-based information to decision-makers and politicians, raising the level of awareness of the general public, popularizing existing expert knowledge, and inspiring individuals with an exhortation to take ownership of these ideas and expand on them. If this book serves to bring about these developments in even a minor way, it will not have been written in vain.

While writing this book, I was, together with my very patient family, literally living with its subject matter. From the start of this project, my wife Kirsten Drotner, my daughter Julie Drotner Mouritsen, and my son Jonas Drotner Mouritsen have been wonderful and understanding with regard to this obsession. Along with me, they quickly became enthralled by seaweeds and their culinary possibilities. On an almost daily basis, we ate our way through one species after another and Kirsten started right away to use seaweeds in baking bread and in fish dishes. She was also more than gracious when I redirected our holiday plans so that we could chase after seaweeds or sail in Japan with the local seaweed fishers. A very large and loving thank you to Kirsten for this. I also owe special thanks to Kirsten and Julie for several of the recipes in the book and for a very thorough and critical reading of several versions of the original Danish manuscript. Many of our friends have, with sympathetic insight, shared my interest in seaweeds. I am grateful to Tina and John Gillis for their hospitality on Gotts Island, Maine, and for fearlessly making room for my homemade 'laverbread' on the breakfast table.

Of the many individuals who have freely put their knowledge and expertise regarding scientific and technical matters at my disposal, particular thanks are due to:—Dr. Susse Wegeberg, who enthusiastically helped me to master the biology of marine algae and who, in addition, critically reviewed parts of the manuscript;—my good colleague the chemist Carl Th. Pedersen for his meticulous perusal of the chemistry underlying the gastronomy and for drawings of several of the chemical structures;—Professor Louis Druehl, an expert on brown algae, for taking me along on the seaweed harvest off Bamfield on Vancouver Island, Canada, and for passing on to me his enthusiasm for seaweeds, both wild and cultivated;—Rae Hopkins for showing me Canadian Kelp Resources' facilities for drying and packaging seaweeds;—*nori* fisher Norio Kinman, marine algae researcher Dr. Norio Kikuchi of the Coastal Branch of the Natural History Museum and Institute, Chiba, Japan, and Dr. Yuichi Kotani, Sekei National Fisheries Research Institute, Nagasaki, Japan, for showing me around the places where *Porphyra* is cultivated in the Chiba and Saga prefectures;—Shep Erhart for his hospitality and tour of Maine Coast Sea Vegetables in Franklin, Maine;—the seaweed experts Drs. Norishige Yotsukura, Philippe Potin, Alan T. Critchley, Stefan Kraan, Susan Holdt, Prannie Rhatigan, and Jane Teas for valuable discussions on seaweeds and their uses for human consumption;—Professor Gerhard Jahreis, Friedrich Schiller University in Jena, Germany, for the analysis of the fat content of *Palmaria*;—the seaweed lover Thorkil Degn Johansson for sharing his enthusiasm for edible algae with me and for introducing me to dulse and sea palm from Mendocino, California;—Donal Hickey, director of Arramara Teoranta in Connemara County, Ireland, for a guided tour of the seaweed factory and seaweed manager Dara Flagherty for a demonstration of how knotted wrack is pulled ashore;—the seaweed expert Brian Rudolph for showing me around CP Kelco and informing me about the production of the food additive carrageenan in Denmark;—Ruth Nielsen, curator of the herbarium for spore-bearing plants at the University of Copenhagen, for her tour of, and information about, the collections;—Jenny Bryant, former algae curator at the Natural History Museum in London, England, for guiding me around and giving me access to the collections;—Brian, Alyson, and Ashley Jones from Selwyn's Penclawdd Seafoods in Wales for the tour of the establishment and information on the production of Welsh laverbread;—Thormar Thorbergsson for discussions about chocolate and seaweeds;—Hans Porse and Torben Hee for discussions about Danish seaweed production;—Sjúrður Fróði Olsen and Hans Samuelson for historical information about the uses of seaweeds on the Faroe Islands;—Ágúst Georgsson and Már Jónsson for valuable information about marine algae as food on Iceland;—Henrik Jespersen for information on seaweeds and macroalgae in Norway and for pointing out the poem about seaweeds in Peter Dass' *The Trumpet of Nordland*;—the Egyptologist Paul John Fransen for clarification on the background of the Ebers Papyrus and its alleged reference to the use of seaweeds for medicinal purposes in ancient Egypt; —the biologist Marianne Holmer for information about marine aquaculture;—Anders Fischer, Lars Bisgaard, Søren Andersen, and Kaare Lund Rasmussen for background on the cultural history and ar-chaeological circumstances related to seaweeds;—Eyjolfur Freðgeirsson for stories about seaweed uses in Iceland;—Gerald T. Boalch for information on the collection of seaweeds in England in the Victorian era;—Jens Trudsø for information on the history of the use of carrageenans;—Ines Goñi Alonso and Borja Glez for translating recipes and texts from Spanish;—the chefs Claus Meyer, Jakob Mielcke, Klavs Styrbæk, Torsten Vildgaard, Søren Westh, and Lars Williams for insightful discussions concerning the role of seaweeds in gastronomy.

At a very practical level, this book could not have been produced without the assistance of many others, among whom I would like to thank especially:—the chefs Klavs Styrbæk, Enaitz Ladaburu, Josean Alija, Lars Williams, and René Redzepi, who allowed me to reproduce recipes and photographs of their creations;—the seaweed harvesters and producers Rasmus Bjerregaard, Louis Druehl, Shep Erhart, Eyjólfur Friðgeirsson, Jens Møller, and Símon Sturleson for providing me with samples of their products;—Dr. Alan T. Critchley for providing samples and for useful comments on the manuscript;– the many individuals, organizations, and companies listed at the back of the book who so graciously allowed me to use their pictures and photographs in this volume, in particular Jacob Thue for his professional photography at the herbarium in the Natural History Museum in London.

I would like to express my sincere thanks to my good friend and collab-orator on this book project, Mariela Johansen, who volunteered to translate the book and with whom I have enjoyed numerous conversations on sea-weeds and cooking. She did an admirably professional job in translating the book into English and adapting it for a wider audience. She suggested valu-able revisions to the text in many places, scrutinized it for consistency, and performed elaborate checks of historical and technical facts, as well as the anecdotal evidence. In addition, she helped to decide on the final subtitle of the book. In many ways, the book greatly improved in her hands.

Last but not least, I owe a debt of gratitude to my son, Jonas Drotner Mouritsen, for taking photographs of marine algae and seaweed dishes, for his fastidiously executed drawings of the many different species of seaweeds, and for the exceptional patience and professionalism he displayed throughout the duration of the project in the integration of text, graphics, and design.

Odense, December 2012, OGM

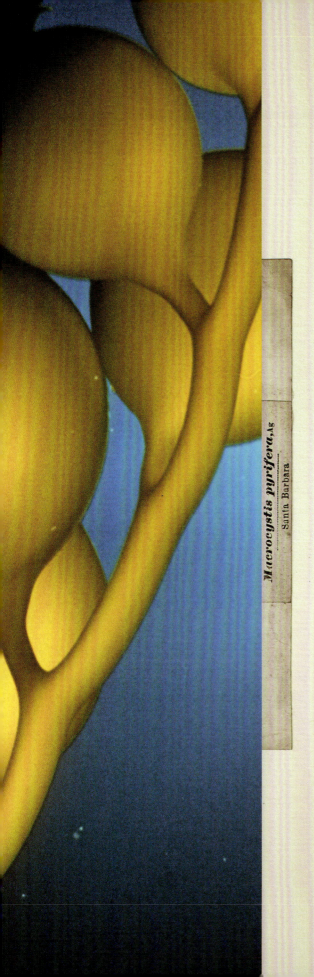

Macrocystis pyrifera, Ag
Santa Barbara

Seaweeds are marine algae

What are seaweeds and marine algae and where do we find them?

AT THE INTERFACE BETWEEN LAND AND SEA

Seaweeds are marine algae

Most people think of seaweeds simply as the plant-like stuff that washes up on the seashore. On a beach in northern Europe, one typically might find a mixture of bladder wrack (*Fucus vesiculosus*) and a variety of seagrasses, such as eelgrass (*Zostera marina*). But for the biologist, bladder wrack and eelgrass are completely different organisms, almost as distinct as plants and animals. In the biological sense of the word, a true seaweed is actually a so-called alga.

The word seaweeds is the popular term that is loosely applied to the larger, more complex marine algae, also called macroalgae. Because all seaweeds are marine algae, the two terms will be used interchangeably in this book.

The word algae is used to designate a large, varied, and heterogeneous group of organisms that, at present, do not have a clear-cut, formal taxonomic status. Some scientists have estimated that there might be between one and ten million different species, by far the majority of which have not yet been described. Just like plants, algae carry out photosynthesis, using sunlight to produce carbohydrates and energy. Of the 35,000 or more currently known species of algae, about half are aquatic, while the others are terrestrial. The aquatic algae are found in fresh and in salt water; it is the latter type, referred to as marine algae, with which we are concerned in this book.

Dr. William Turner (1508–1568), considered the 'father of English botany', is responsible for the linguistic association between seaweeds and plants. When he compiled the first scientific treatise on plants giving them English names in the 16th Century, he regarded seaweeds as useful herbs and included them, using the term 'seawrake'. This was an allusion to their origin, like shipwrecks, as something that washed up from the sea.

Algae come in many different sizes. The smallest of them, the microalgae, are unicellular and make up what we call plant (or phyto) plankton. Some of them are related to animal plankton, bacteria, and fungi. The largest algae are multicellular organisms, growing to lengths of up to 60 meters, which can form enormous 'forests' in the ocean. These large marine algae, which are also referred to as macroalgae, are the ones that most people associate with the word seaweeds.

Seaweeds are found in all coastal areas of the world, in all climatic zones from the warm tropics to the icy polar regions. There are about 10,000 different species, but new and formerly unknown ones, sometimes living under extremely harsh conditions, are being discovered on an on-going basis. Fossil finds have shown that seaweeds are a form of life going back at least 500 million years and that algae, of one type or another, have existed on Earth for about three billion years. There is also much evidence that they have not changed significantly during this time.

Despite their name and even though they often resemble plants, seaweeds are only tenuously related to them. The tissue of the majority of seaweeds is built up very differently from that found in higher forms of plant life and their

◄ Underwater seaweed forest.

What are seaweeds and marine algae and where do we find them?

functional structure is dissimilar in many respects. They do not have leaves and stems in the botanical sense of the words, nor do they bloom, produce seeds, or set fruit. Seaweeds have no need for a root system to take in water or nutrients, as their cells are in direct contact with the surrounding water from which they derive their nourishment. Consequently, they do not grow roots. Some species have evolved a system for the internal transport of vital salts and the products of photosynthesis, but the majority are undifferentiated, with each cell being responsible for generating what it needs.

Seaweeds are to the sea what forests, undergrowth, bushes, and groundcover are on the land. They produce oxygen and release it into their surroundings while at the same time functioning as a physical structure that provides a habitat for a wealth of other organisms. Because they do not need roots, a few species, such as sea lettuce (*Ulva lactuca*) and gulfweed (*Sargassum*), thrive as completely free-floating entities not unlike plankton. A notable example of this is found in the Sargasso Sea, which is home to an independent seaweed-based ecosystem far from any shore. The majority of marine algae, however, find their home in a transition area between the land and the sea, anchoring themselves to firm substrates, such as rock faces and stones, or to the seabed. Since all seaweed species need light in order to thrive, their distribution is determined by local light conditions and water turbidity.

SEAWEEDS OF ALL SIZES, GROWING UNDER ALL CONDITIONS

In this book we will focus on those marine macroalgae that are multicellular organisms. It is noteworthy that some species of seaweeds are very small,

By custom and tradition, some seagrasses have been thought of as seaweeds, even though they are not algae, an example being eelgrass (*Zostera marina*), which is actually a true plant that grows partially submerged. Many also characterize glasswort (*Salicornia europaea*), sometimes called samphire or sea asparagus, which grows in salt water marshes, as a seaweed, but it, too, is a plant. Just to add to the confusion, lichens can also be mistaken for seaweeds. Lichens are actually small ecosystems consisting, for example, of green microalgae in a symbiotic partnership with fungi. This allows the two species to survive in areas where neither could survive on its own.

"In the morning found such abundance of weeds that the ocean seemed to be covered with them; they came from the west." Description of the Sargasso Sea from the log of Christopher Columbus' first voyage to the New World, September 21, 1492.

▸ Seaweed-covered rocks on the coast of Maine, USA.

only a few millimeters or centimeters in size, while others are gigantic. One species, the giant kelp (*Macrocystis pyrifera*), can regularly attain lengths of 60 meters and form enormous 'kelp forests' in the ocean. A large number of seaweed species have adapted to life in the intertidal zone, where twice daily they are exposed to the air. Others can survive only if they are constantly under water. It is estimated that seaweeds take up an area that corresponds to about 8% of the total area covered by the world's oceans.

Many species of seaweeds are surprisingly robust; some tolerate being dried out completely, being exposed to frost, or being subjected to great fluctuations in temperature. They endure the hardships caused by rapid ocean currents, violent tidal changes, foaming surf, and mighty waves pounding against cliffs and coastlines. Others are able to withstand considerable variations in salt concentrations, ranging from the open sea to areas of brackish water.

SEAWEEDS COME IN MANY COLORS

Traditionally, seaweeds are divided into three main groups: green algae, red algae, and brown algae. Even though this classification is unambiguous, one cannot always use color to determine how to classify a given seaweed species. This characteristic varies with the number and types of pigments it contains, as well as its tissue structure.

All species of seaweeds have chlorophyll granules that contain chlorophyll *a*. This substance is green and is part of the photosynthetic system, which converts sunlight to the chemical energy that fuels the metabolic functions of the seaweed. Nevertheless, its green color is often masked by a number of other pigments, resulting in brown, yellowish, and red tones. The color of green algae is overwhelmingly due to chlorophyll *a*. In the red algae, certain other pigments, called phycobilins, impart red, orange, and blue hues. Brown algae contain only a little chlorophyll and their brownish-yellow color is due to a pigment called fucoxanthin. A similar brownish pigment is found in those plants that take on the familiar red, yellow, and brown autumn colors when their otherwise dominant green chlorophyll *a* disappears.

Because phycobilins are water soluble, seaweeds, particularly the red algae, often lose some of their color when they are pried loose from the place where they are growing and set adrift in the sea. The green color fades more slowly because chlorophyll is insoluble in water.

Seaweeds throughout the ages

SEAWEEDS AND HUMAN EVOLUTION

One of the characteristics of humans as a species is that our brains are large in proportion to our body mass. How did this come about? It is now generally acknowledged that the ancestors of present-day humans, the upright primates, did not evolve on dry, warm grasslands but in the damp, warm regions that formed the border between land and water. The British neurochemist Michael Crawford has pointed out that the all-important sources of essential and superunsaturated omega-3 fatty acids, especially DHA (docosahexaenoic acid) and EPA (eicosapentaenoic acid), can be found in sufficient quantities only in littoral areas, where fish and shellfish are abundant. These fatty acids, with which we are also familiar from fish oil and food supplements, are a vital requirement for the formation of a complex nervous system and a large brain. Hence, it was a determining factor in the evolution of modern humans that our ancestors, during the period from about 1,000,000 years ago until the appearance of the first modern *Homo sapiens* about 100,000–200,000 years ago, had a diet that consisted of fish and shellfish.

It is, nevertheless, more difficult to determine conclusively whether seaweeds were also incorporated into the diet of our distant ancestors. Seaweeds are, however, a primary source of omega-3 fatty acids and it is actually from them and other algae that marine animals derive these substances. Studies of fossil finds indicate that the enamel surface of the teeth of early hominids shows evidence of wear that is characteristic of eating food containing a certain content of silica particles, which are typically found in wetland plants. In addition, analysis of the minerals found in fossilized bones has shown

▲ Seaweeds grow in many shapes and colors.

that the proportion of strontium to calcium is so low as to suggest that these individuals were not very high up in the food chain and, consequently, were probably herbivores. So it is reasonable to suppose that seaweeds, given their prevalence in coastal areas, also played a role in the early hominid diet.

Seaweeds are marine algae

But there is another basis for conjecturing that hominids evolved in coastal areas. The Canadian researcher Stephen Cunnane, through his research on the relationship between evolution and nutrition, has drawn attention to the fact that the development of the brain is especially dependent on five micronutrients: iodine, iron, copper, zinc, and selenium. Without a sufficient supply of these, the genetic potential for the evolution of a large and complex brain could not be maximized. Iodine plays a particular role because it is a prerequisite for the production, in the thyroid gland, of two important iodinated hormones, thyroxine and triiodothyronine, which regulate metabolism and control growth rate. To date, there is only scant knowledge of precisely how iodine deficiency affects the development of the brain. It is Cunnane's hypothesis that hominids living far from the sea would have been unable to obtain enough iodine to evolve a brain such as that found in humans today.

People who live in coastal areas rarely have diseases attributed to lack of iodine. It is also noteworthy that no wild animals, including primates, exhibit signs of iodine deficiency. Seemingly, their bodies are in equilibrium with a diet that suits the requirements of their genetic makeup.

We now know that another effect of iodine deficiency in humans is goitre, which results in an enlarged thyroid gland and a swollen neck. About two hundred years ago, the close connection between iodine on the one side and seafood and seaweeds on the other side, especially the brown algae that contain much larger quantities of iodine than terrestrial plants, gained acceptance as a scientific fact. For close to a century, many countries have made it compulsory to add iodine to table salt as a public health precaution. A number of species of brown algae, commonly referred to as kelp, which are a good source of both salt and iodine, have for a long time been used in the production of table salt to achieve this effect.

SEAWEEDS IN A HISTORICAL PERSPECTIVE

Seaweeds break down easily and, as a result, there is no conclusive archaeological proof that humans used seaweeds as food in prehistoric times. But recent discoveries are indicative of the use of marine algae by coastal peoples as long ago as 12,000 BCE. At an excavation in Monte Verde in southern Chile, remains of several seaweed species dating from this period have been found in a mortar and in hearths. These samples were preserved because the site is located in what is now a peat bog where the acidity of the soil inhibited bacterial decay. While it is difficult to determine whether the seaweed remnants were used in a sort of bread or cake or for medicinal purposes, it is unlikely that seaweeds were simply burned for fuel.

The earliest known medicinal uses of seaweeds can be traced to the principal schools of classical medicine in China, Japan, and India (ayurveda). The legendary emperor Shen Nung, who lived almost 5,000 years ago and who is also said to have discovered tea, is credited with founding traditional Chinese herbal medicine by exploiting the therapeutic effects of plants and seaweeds. According to the emperor, seaweeds, because of their salty and spicy taste, could be used to soften, moisten, and soothe, as well as to relieve muscle tension and to shrink lumps and tumors. Their use was, therefore, recommended for curing constipation, endocrine diseases, cysts, and chronic bronchitis. Unfortunately, the original text of Shen Nung's *Materia Medica* was lost and portions of it were first recorded in a secondary written source several thousand years after his death. In another early work on Chinese medicine, from ca. 300 BCE, Chi Han identifies seaweed extracts as an insect repellent.

With regard to the treatment of goitre specifically, it is possible that people living in coastal areas had some folkloric knowledge that eating iodine-rich foods was a way to prevent this disease. The emperor Shen Nung recommended the use of *Sargassum* and, in the 4th Century, Ghe-Khun identified *Sargassum siliquastrum* and *Saccharina japonica* in an alcohol solution as effective remedies. The earliest recorded reference to the use of seaweeds in Western medicine dates from ca. 1200 CE when Ruggiero Frugardi and his pupil, Rolando Capulletti, of the eminent medical school at Salerno, explicitly noted the beneficial effects of using dried or burned seaweeds to counteract goitre. In the mid-1700's, the English doctor Bernard Russell used burned brown algae both internally and topically to treat cases of goitre and scrofula.

Claims have been made that there are references to seaweeds in the Egyptian Ebers Papyrus, which dates from about 1500 BCE and is the oldest existing document on the art of healing. In it, plants are described as curatives for boils and tumors. While there have been discussions about whether a description in the Ebers Papyrus can plausibly be interpreted as a treatment for breast cancer using seaweeds, recent examinations of the text seem to give no grounds for doing so.

Undoubtedly many different species of seaweeds have been used as food in coastal regions all over the world for thousands of years. Nevertheless, it is noteworthy that the exploitation of this resource has been very limited in Europe and North America in comparison with Asia, Polynesia, South America,

Australia, and New Zealand. In Southeast Asia and Polynesia, in particular, where seaweeds were prized both by the ordinary people and the nobility, they have preserved their status as an important foodstuff right up to the present. Edible seaweeds have played an especially significant role in Japan, thanks to its long coastline and the wealth of seaweed species to be found there.

Seaweeds are marine algae

In Japan, China, and Korea, marine algae have been treated with reverence and respect and in earlier times were associated with wellness. There were even periods when the best seaweeds were reserved for the nobility. They are often mentioned in classical Japanese poetry from the 6th and 7th Centuries, where one finds verses about the fishermen's wives who collect seaweeds for salt production and for the payment of taxes. The oldest Japanese-Chinese dictionary, dated 934 CE, describes 21 different species of edible seaweeds and gives instructions for their preparation. Many of these recipes have survived to the present day. In Japanese *Shinto* temples, the burning of seaweeds and the offering of seaweeds and other foodstuffs to the gods has played a major role as a fertility symbol and as a prayer for an abundant harvest.

The particular esteem with which the Japanese regard seaweeds is mirrored in the yearly Seaweed Day, celebrated on the 6th of February. On that day in 701, seaweeds, especially *nori* and *arame*, were inscribed on the list of valuable goods that could be offered to the emperor. At the time, this tax was a serious burden for the population in Japanese coastal areas, as they had to make payments to the imperial court, to the military authorities, and to the temples. Seaweeds continued to be used as tax contributions until the 18th Century.

For thousands of years, seaweeds have formed an important element of the diet in the Hawaiian and other Polynesian islands. Called *limu*, they were cultivated in special sea gardens, with more than 70 different species being used either as food, as medicine, or for religious ceremonies. On Hawaii alone, at least 40 species were eaten raw, baked, pickled, or mixed with other foodstuffs. The Polynesians brought this custom to New Zealand where the Maoris ate red algae (*karengo*) and used brown algae for the preservation of sea birds.

The technical application of seaweed extracts as thickeners and gelling agents goes back several centuries. The methodology for extracting a gelling agent, known as agar, from seaweeds was first described in 1658 in China and shortly thereafter introduced to Japan, where agar goes by the name of *kanten*. We also know agar as a substrate for bacterial culture, a method introduced by the Nobel laureate Robert Koch in 1881, on the suggestion of Fannie Hesse, the wife of a colleague who used seaweed extracts to thicken jellies.

Was there a seaweed route to South America?

The peopling of North America has for decades been a much discussed subject—where did the first inhabitants come from and when did they arrive? The widely accepted answer now is that the first North Americans came from Asia at least 14,000 years ago. At that time, what is now Alaska was still connected to eastern Siberia by a broad grassland steppe, Beringia. This theory has been validated by the most recent analysis of DNA samples extracted from dried human excrement found in a cave in Oregon.

Of equal importance is the question of which route these early inhabitants followed as they gradually spread southward to the very tip of the South American continent. The recent discovery and examination of archaeological remains of seaweeds found in a hearth excavated at Monte Verde in southern Chile seem to hold the key to the puzzle.

Since the late 1970's, Tom Dillehay, from Vanderbilt University in the United States, and his Chilean colleagues have analyzed findings from this site and suggest that it is now possible to conclude that the migration proceeded along the coastlines. This migration path has been called 'the kelp highway'. The discovery of ample remains of so many different species of seaweeds shows that the people living there had a detailed understanding of how to use about twenty different marine macroalgae for food and medicine. Among the seaweed genera found were *Porphyra, Gracilaria, Sargassum, Macrocystis,* and *Durvillaea.*

As Monte Verde is located far inland, the fairly advanced exploitation of marine algae implies that the population must, in an earlier period, have acquired significant knowledge about finding food along the seashore at different times of year and in a variety of coastal settings. This seems to indicate that this part of the continent was settled by people who had arrived from the north by following the shoreline rather than by moving overland. Traces of seaweeds found on stone tools also show that the seaweeds were worked in some fashion, possibly by chopping them into pieces or by grinding them for medicinal purposes.

Interestingly enough, some of the seaweed species found at Monte Verde are still used by the local population to treat common health problems.

▲ Microscopic view of remains of the brown alga *Sargassum* sp. found on the floor of a hut at Monte Verde in southern Chile, dating from about 12,000 BCE.

▲ Stone tool from Monte Verde with remnants of seaweed along its edge.

Seaweeds in Europe

In Europe, it is especially Ireland, Wales, Scotland, Brittany, Iceland, and Norway that have historical connections with seaweeds, which throughout the ages have been an important socio-economic factor in these areas.

Seaweeds are marine algae

Stories are told of Irish monks of the 12th Century gathering dulse (*Palmaria palmata*) for distribution to the poor, who probably ate it for lack of anything better. The Irish used seaweeds in a number of ways: as food, as a sort of chewing tobacco, and for medicinal purposes, e.g., as a cure for worms, for colds, and for women's homesickness! From about 1600 to the present day, Irish people have exploited seaweeds commercially, harvesting them systematically for field fertilizer, the extraction of iodine, the recovery of soda and potash used in soap and glass production, and more recently for technologically innovative products.

Ireland's close connection with seaweeds is literally embedded in the name of a smaller species of red alga called carrageen (*Chondrus crispus*, known popularly in English as Irish moss), a name derived from the Irish word *carraigin* (little rock). It is found not only on the shores of the Emerald Isle, but also along coastlines in Europe and North America. As indicated by the name, the seaweed is an excellent source of carrageenan, one of the substances used as a gelling agent. In spite of the long tradition of exploiting marine algae in Ireland, most people, unfortunately, associate them with poverty and the starvation that resulted from the Great Famine that devastated the country in 1847–48.

On the other hand, in Wales no stigma is attached to the use of seaweeds as food and purple laver (*Porphyra umbilicalis*) is still eaten in salads, biscuits, and as an accompaniment to roasted meat.

In Brittany the use of seaweeds as human food goes back at least as far as it does in Ireland and the vernacular names for dulse (*Palmaria palmata*), *dillisk* in Ireland and *tellesk* in Brittany, are indicative of a common origin. The centuries old practice of using dulse for animal fodder continues to the present. Here, as in other parts of France, various types of brown algae have been used to fertilize the fields since the 1300's and, more recently, seaweeds have also been used as fertilizer in vineyards. Detailed regulations governing who had the right to collect and harvest seaweeds suggest that at some points in history it had considerable socio-economic significance.

Potash derived from seaweeds was used in glass making in Europe throughout the Middle Ages right up to the beginning of the 19th Century.

▸▸ Diagrammatic illustrations of red algae from Ernst Heinrich Philipp August Haeckel's classical natural history monograph, *Kunstformen der natur* (Artforms in Nature), 1904.

▶ Collecting seaweeds on the coast of Brittany.

Soda used in glass making can be made from glasswort (*Salicornia europaea*). It is said that, although it grew abundantly, glasswort had no special name in English prior to the arrival of Venetian glassmakers in England in the 16th Century. They immediately recognized its potential end use and the marine plant was named accordingly.

A building material combining clay with seaweeds, which add structure and binding power, has been used to construct both houses and dikes. It is said that Venice is built of poles, mud, and seaweeds.

It was, in fact, one of the secrets of the famous Venetian glassblowers on the island of Murano. It enabled them to create exceptionally clear glass that could be kept malleable for a longer period of time, allowing it to be fashioned into elaborate shapes.

The common designation for some brown seaweeds is kelp, derived from the French word *culpe*, a medieval term that was actually used more to denote seaweed ash than seaweeds. Around the beginning of the 1800's, the French started to extract iodine from seaweed ash. This led to a significant industrial development, but one that had its ups and downs. At its height in the 1930's, there were almost 3,000 small boats, especially in Brittany, engaged in harvesting mainly brown algae of the order Laminariales for iodine production. These boats collected more than 200,000 tons of fresh seaweeds yearly. After World War II, the need for iodine declined and the French seaweed industry turned to the production of alginates, which are used as thickeners. With modern optimal harvesting methods, only a couple of dozen boats are now engaged in the business, collecting about 50,000 tons of fresh algae each year.

SEAWEEDS IN THE NORDIC COUNTRIES

An assortment of anchored seaweed species are found along all the coastlines of the European Nordic countries, from Kattegat, between Denmark and Sweden, in the south to the Arctic Ocean in the north. Going back to the earliest times of human habitation in these areas, there are indications that marine algae have been used in a variety of ways, as food for people and animals and as a material that was eminently suitable for a number of practical, everyday purposes.

Collecting seaweeds in Victorian England

In the first half of the 19th Century, English laymen developed a growing interest in, and infatuation with, natural history, leading them to devote their leisure time to the independent study of Nature, often by the seashore. Not long before, the Swedish naturalist Carl von Linné had introduced a classification system, but many species had not yet been designated by unique Latin names and were known by names that varied from one district to another. English ladies, in particular, enjoyed gathering up strange objects on the beach. Often led by the local curate or doctor, they went along the shore at low tide collecting bits of seaweed which they dried, pressed carefully, and mounted on paper. These specimens were labelled with fanciful names such as mermaid's shaving brush, sea-girdles, peacock's tail, and sea lace.

The interest in seaweeds seems also to have inspired poets to point out the injustice inherent in referring to them as weeds. In the little poem below, the seaweeds themselves ask to be called "Ocean's gay flowers". Today many seaweed lovers prefer to call them vegetables from the sea.

Algae, bright order!
By Cryptogamists defended –
Translate marine plants
as Linnaeus intended.
You collect and admire us,
we amuse leisure hours;
"Then call us not weeds,
we are Ocean's gay flowers."
(Curtis)

◄ A folded book with pressed seaweed samples, probably collected by a girl on one of the English Channel Islands at the beginning of the 19th Century. The small verse quoted here is inscribed on the back page. Cryptogamia refers to an archaic botanical classification system.

Seaweeds are marine algae

▶ Caricature of the collection of seaweeds and other strange objects on an English beach. Drawing by John Leech from the satirical magazine *Punch*, 1856.

▶▶ Detail from the oil painting *Loading seaweeds at Hornbæk Beach* (1882) by the Danish artist Carl Locher (1851–1915).

As the last glacial period in Scandinavia came to an end, between about 12,000 and 8900 BCE, people moved north and established themselves in the coastal areas. Meal scraps found in these late Paleolithic settlements contain remnants of bivalve shells, fish bones, and the bones of marine mammals, such as whales and seals. The evidence from these middens indicates that the population was overwhelmingly dependent on, and closely bound to, the sea as a food source. It is impossible, however, to know whether the early Scandinavians supplemented their diet with seaweeds, as algae leave behind no clear traces in the way that shells and bones do. There is no proof that the charred bits of seaweed that have been found either in hearths or in earthenware vessels were meant for human consumption, because the seaweeds could just as easily have been used as fuel for the fire. Nevertheless, it seems highly probable that such a readily available food source would have played a role in the late Stone Age diet before agriculture took hold and there was easier access to plant foodstuffs.

Among Nordic countries, Iceland has the best documented use of seaweeds in historical times. It is possible that the idea of eating seaweeds came to Iceland and the Faroe Islands from England and Scotland and then migrated from Iceland to Norway, where the consumption of dulse, in particular, was widespread in coastal areas. In all of these regions, it was easy to gather dulse along the low-lying shores and dry it for later use. Dulse was used as a trading commodity on Iceland since the 700's and was one of the goods exchanged between coastal and inland dwellers. It was effectively a hard

currency; the cost of renting a farm was often expressed in terms of a quantity of seaweeds. The seaweeds were collected at the end of June by gatherers who lived in tents pitched along the beach for the duration of the harvest. People came to the seashore to barter for seaweeds—a kilogram of dried fish could be exchanged for a kilogram of dulse. The harvested dulse was rinsed in water and then spread out to dry. In the course of the desiccation process, salts and amino acids sometimes seep out to form a layer on the surface of the seaweeds. The greater the deposit of this salty-sweet powder, the more the seaweeds were appreciated.

Seaweeds are marine algae

Written sources, namely sagas and law codes, record the use of seaweeds as human food on Iceland as far back as the 10th Century. A number of different types of seaweeds were harvested, in particular red algae. Icelanders, and possibly also Norwegians, ate fresh dulse baked in bread and dried and salted dulse as a sort of snack. For the preparation of a meal, the seaweed was mixed with butter or lard and served with dried fish or cooked potatoes and turnips. Another way of preparing dulse was to cook it with milk or put it in porridge. Finally, dulse was added to bread dough in order to make the flour stretch farther. The Norwegian Vikings undoubtedly brought dried seaweeds with them as provisions for their long expeditions.

Throughout the Middle Ages, dulse (*Palmaria palmata*) was incorporated into the Icelandic diet. A written source from 1555 cites it as an obligatory component of the meals eaten by the pupils at the Latin (grammar) school.

During the winter, the Inuit of Greenland ate cooked seaweeds (winged kelp, bladder wrack, dulse, and knotted wrack), which, outside of the hunting season, could serve as important sources of vitamin C. On Greenland one can still find older people who eat winged kelp in soups. In Tasiilaq, a town in eastern Greenland, there is an enduring tradition of going on marine outings to gather snails and mussels, which are boiled and eaten with fresh sugar kelp (*Saccharina latissima*) that has been collected at low tide.

When food was in short supply and there was a risk of famine, the Icelanders probably also had to resort to species that were hardier than dulse. The most probable candidates were carrageen (*Chondrus crispus*) and winged kelp (*Alaria esculenta*), which were normally used only for animal fodder. The consumption of seaweeds was, however, for many centuries linked mostly to times of hardship. On the Faroe Islands there is even a saying about a person who dies in poverty: "He was laid in his grave with a piece of seaweed in his mouth."

In earlier times, seaweeds and seaweed meal were also used as fodder for domestic animals such as cows, goats, and sheep, especially in the coastal regions of Norway and on Iceland and the Faroe Islands. The value of seaweeds as fodder for cattle is highlighted in a famous Norwegian verse from *The Trumpet of Nordland*, written by the poet-priest Peter Dass toward the end of the 17th Century. In it he tells of a poor fisherman living on a small island. His only livestock was single cow. The animal had nothing to eat but seaweeds and the leaves of briny vetch, but it grew much fatter than any that grazed in a meadow.

In the years when the grass grew poorly and the hay harvest was a meagre one, seaweeds, with their considerable nutritional value, could be a supplement. Domestic animals either found the beached seaweeds on their own or else people gathered the dried clusters and poured boiling water over them before they were used as fodder. On occasion, the softened seaweeds were also mixed with fish scraps and hay. Here again, dulse was the preferred species, but other, less delicate, types of seaweeds were also collected. On Iceland, the nutritional value of the fodder was improved by letting the damp seaweeds ferment in a hole in the ground over the winter.

Since ancient times, seaweeds have probably also been used for the production of salt. When they are burned there is a residue of ashes containing salts that are then leached out in seawater. When the water is evaporated, the result is a black salt that contains some of the combustion products from the seaweeds. This method of production was utilized in many coastal areas in the Nordic countries, being especially important during periods when there was a shortage of imported white sea salt, used for preserving fresh foodstuffs.

Throughout the Middle Ages, and in Norway right up to the 1900's, seaweeds, particularly bladder wrack (*Fucus vesiculosus*), and eelgrass (*Zostera marina*) have been used as fertilizer on the fields, often in combination with animal manure. Seaweeds contain phosphorous, nitrogen, and especially potassium in reasonable quantities, but their salt content makes them less than ideal for use on a number of crops.

At times in Nordic countries, seaweeds were used for the production of soda, formerly an important component in the manufacture of glass, glazes, and soap. In many places, the soda production was actually based on plants from the intertidal zone, an example being glasswort or saltwort, whose name mirrors its end use.

On account of their salt content, dried seaweeds and sea grasses are resistant to decay, fungus infestation, and insect attack. Dried eelgrass (*Zostera marina*) was used as a sturdy and durable stuffing in mattresses, pillows, and furniture cushions. In fact, seaweed mattresses were commonly found in some parts of Denmark right up until the middle of the 20th Century. These same properties made dried seaweeds useful for sea dikes and as fences between

Seaweeds in an Icelandic saga

In the Saga of Egil Skallagrimsson, which dates from the 10th Century, seaweeds (in this case, the red alga dulse) are described in such a way that one gains the impression that they were something that could impart renewed vitality and love of life. This section of the saga deals with Egil's sorrow after he had placed the body of his beloved son Bødvar, who had drowned, in the burial mound of the Skallagrim family.

Then Egil rode home to Borg, and when he arrived he went at once to the locked bed-closet in which he usually slept, lay down, and shut himself in. No one dared speak to him. It is said that when they laid Bødvar to rest Egil was wearing laced stockings and a tight-fitting tunic of red fustian, laced at the side; but they say that his muscles so swelled with the exertion that both the tunic and the stockings tore open. The next day Egil still did not open the bed-closet and took neither food nor drink. He lay there that whole day and the following night, no one daring to speak to him. On the third morning, as soon as it was light, Asgerd sent a man on horseback to ride as fast and hard as he could westward to Hjardarholt to ask Thorgerd to come to Borg without delay. The messenger arrived in the afternoon and Thorgerd at once asked for her horse to be saddled. Together with two men who attended her, she rode that evening and through the night until they came to Borg. Thorgerd at once entered the kitchen hall where the fire was burning and Asgerd greeted her, asking whether they had eaten. "No, I have not," Thorgerd replied loudly, "and I have no intention of taking any food in my mouth until I sup with Freya. I know of no better counsel than that of my father—I will not outlive my father and my brother." So she went to the bed-closet and called out, "Father, open the door! It is my wish that we should accompany each other." Egil undid the lock; Thorgerd stepped up into the bed-closet, locked the door behind her, and lay down on one of the other beds. Then Egil said, "It is pleasing, my daughter, that you are following along with your father. You have shown me much affection. How can anyone

expect that I have the will to live after such a misfortune?" After this they were silent for a while. Then Egil spoke again, "What is going on, my daughter—are you chewing on something?" "I am chewing seaweed," replied Thorgerd, "because I believe it is bad for me. Otherwise death will never come." "Is it then harmful?" asked Egil. "Very bad," she replied. "Do you want some?" "Why not," he said. A little while later she called out and asked for something to drink and water was brought to her in a drinking horn. "That's what happens when one eats seaweeds," said Egil, "one becomes even more thirsty." "Would you like something to drink, too?" she asked him. He took the drinking horn and drank a great quantity. "We have been fooled," said Thorgerd, "they gave us milk." On hearing this Egil bit a shard out of the horn reaching as far down as he could with his teeth and threw down the horn. Then Thorgerd asked, "What shall we do now? Our intentions have been thwarted. I think, father, that we might as well live on at least until you have composed a memorial poem for Bødvar. Then we can die if we want to."

And so it was that Egil composed his famous funeral verse, *The Loss of Sons.*

Seaweeds are marine algae

▸ Eelgrass (*Zostera marina*) used as roofing material on a house on the island of Læsø, Denmark.

Archaeological findings in Denmark prove that dried seaweeds and eelgrass were utilized to build up burial mounds near the coast and as protective wrapping for urns and chests in Bronze Age graves.

fields, as well as in construction. Seaweeds have the additional advantages that they are fire resistant and act as both heat and sound barriers. As a consequence, they were valuable building materials for roofing and insulation.

The durability of dried seaweeds, as well as their valuable insulating properties, were very important in times when heating and, especially, cooling were either difficult or impossible to achieve. Dried seaweeds were used to insulate ice pits where one placed large blocks of ice, cut from frozen lakes or marshes during the winter, for keeping food cool the following summer. Similarly, this material could be used to protect food stored in root cellars, such as turnips and potatoes for human and animal consumption, from frost and cold.

Although fire resistant, when washed thoroughly to remove as much salt as possible and then dried, seaweeds are combustible and can be used for fuel. On Iceland, for example, bladder wrack (*Fucus vesiculosus*) and knotted wrack (*Ascophyllum nodosum*) were commonly gathered for this purpose. While seaweeds are a very cheap resource, they are not particularly cost-effective on account of their low heat output.

In addition, seaweeds were used for medicinal purposes. For example, bladder wrack (*Fucus vesiculosus*) has been used in Denmark since the 1400's as a headache medicine, sugar kelp (*Saccharina latissima*) was used to alleviate plugged ears, and a number of different types were used as remedies for arthritis and, misleadingly, as a cure for baldness. The Icelanders knew how to apply poultices containing dulse (*Palmaria palmata*) to fight infection, while extracts of marine algae were thought to counter sea sickness and hangovers.

20

In recent times, the seaweed growths along the coasts of Norway have been exploited more extensively than those in other areas. Since the 1970's, the Norwegian seaweed industry, which is based on the harvest of naturally occurring brown algae, especially tangleweed (*Laminaria hyperborea*) and knotted wrack (*Ascophyllum nodosum*), has developed into a significant commercial enterprise. About 150,000 tons of fresh seaweeds are harvested every year. A Norwegian company, FMC BioPolymer A/S, is now the world's second largest producer of seaweed alginates for biomedical and technical end uses. On Iceland, the company Thorverk has harvested knotted wrack and tangle (*Laminaria digitata*) since 1986 for the production of fodder and alginates.

The biology of algae

MICROALGAE AND MACROALGAE

Phycology, the study of algae, is a word derived from the Greek *phycos*, which means seaweed, and *legein*, to speak. The taxonomy of algae is complicated and is still not fully understood. The total number of species is unknown; 35,000 to 50,000 is a typical estimate, but some researchers think that this is a gross underestimate. One speaks of both microalgae and macroalgae.

For the most part, microalgae are microscopic, unicellular organisms and they are divided into a whole range of groups, of which diatoms and the blue-green microalgae are probably the best known. Some microalgae float or swim freely in the ocean, particularly in the upper, light-filled layers, while others fasten on to stones or the surfaces of seaweeds. Diatoms belong to the phytoplankton group (also called plant plankton), which forms the base of the food chain pyramid. Blue-green microalgae are no longered considered algae but photosynthetic bacteria, which can, under certain circumstances, contain toxins. During periods when there is a pronounced algal bloom in the ocean, they can also pose a serious danger to the other living organisms of the sea because dead algae sink to the ocean floor and their decomposition leads to a decrease in the dissolved oxygen content of the water.

Macroalgae are, loosely speaking, those that can be seen with the naked eye. Most of them are classified as 'benthic', which is to say that they fasten themselves to the seabed. Alternatively, they attach to the surfaces of other organisms, such as mussels, but some—for example, sea lettuce (*Ulva lactuca*) and black carrageen (*Furcellaria lumbricalis*)—can grow as completely free-floating entities.

▶ An abundant variety of
seaweeds on the ocean floor off
the coast of France. The large
brown alga, *wakame* (*Undaria
pinnatifida*), is growing along-
side several different species
of red and green seaweeds.

The term 'blue-green algae' is
outdated, but it is still so common that it is used in this book.
Some of these organisms are not
real algae, which are eukaryotes,
but rather bacteria, i.e., prokaryotes, and their correct scientific
classification is 'cyanobacteria'.
They are distinguished from
other bacteria by their ability to
perform photosynthesis. In this
capacity, they are believed to
have changed the atmospheric
conditions on Earth about three
billion years ago, paving the
way for the evolution of higher
organisms that require oxygen.

THE STRUCTURE OF MACROALGAE

This book focuses on those marine benthic macroalgae that we commonly
refer to by the single term seaweeds. Macroalgae are multicellular. Although
some of the larger ones have complex structures with special tissues that
provide support or transport nutrients and the products of photosynthesis,
others are made up of cells that are virtually identical. The smallest seaweeds
are only a few millimeters or centimeters in size, while the largest routinely
grow to a length of 30 to 50 meters. Seaweed cells come in different sizes; in
many species they can measure one centimeter or more. These large cells can
contain several cell nuclei and organelles in order to ensure that the production of proteins is sufficient to sustain the function of the cell and the rapid
growth of the seaweed.

Macroalgae are classified into three major groups: brown algae (*Phaeophyceae*), green algae (*Chlorophyta*), and red algae (*Rhodophyta*). As all of the
groups contain chlorophyll granules, their characteristic colors are derived
from other pigments. Many of the brown algae are referred to simply as kelp.

It is estimated that 1,800 different brown macroalgae, 6,200 red macroalgae, and 1,800 green macroalgae are found in the marine environment.
Although the red algae are more diverse, the brown ones are the largest.
Even though we talk about the three groups of seaweeds as if they were closely
related, this is true only to a minor extent. For example, brown algae and red

algae belong to two different biological kingdoms and are, in a sense, less related to each other than, for example, a jellyfish is to a bony fish. Green algae and red algae are more closely related to higher plants than brown algae are and, together with diatoms, they evolved earlier than brown algae.

Most species of seaweeds have soft tissues but some are, to a greater or lesser degree, calcified, an example being calcareous red algae. The growth of the calcium layer is precisely controlled by the polysaccharides that are present on their cell walls.

Seaweeds, especially the brown algae, are generally made up of three distinctly recognizable parts. At the bottom there is a root-like structure, the holdfast, which, as the name implies, secures the organism to its habitat. It is usually joined by a stipe (or stem) to the leaf-like blades. The seaweed can have one or more blades and the blades can have different shapes. In some cases, the blades have a distinct midrib. Photosynthesis takes place primarily in the blades and it is, therefore, important that the stipe is long enough to place them sufficiently close to the surface of the water to reach the light. Some species have air-filled bladders, a familiar sight on bladder wrack, which ensure their access to light by holding them upright in the water. These bladders can be up to 15 centimeters in diameter. Because brown algae are so much like plants, they are often confused with them.

Some smaller species of seaweeds have a tissue that has a less distinctive structure, consisting only of filaments of cells, which may or may not be branched.

THE LIFE HISTORY AND REPRODUCTIVE PATTERNS OF MACROALGAE

The life history of algae is complicated and this is what really differentiates them from plants. That of macroalgae can consist of several stages that are often so distinct that, in the past, they have caused them to be mistaken for separate species. Reproduction can involve either exclusively sexual or asexual phases, while some species display an alternation of generations that involves both in succession. In the former, the seaweed produces gametes (egg and sperm cells) with a single set of chromosomes and, in the latter, spores containing two sets of chromosomes. Some species can also reproduce asexually by fragmentation, that is, the blades shed small pieces that develop into completely independent organisms.

Asexual reproduction allows for fast propagation of the species but carries with it an inherent danger of limited genetic variation. Sexual reproduction ensures better genetic variation but it leaves the species that depend on this

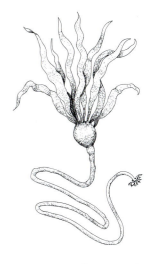

▲ Diagrammatic illustration of a seaweed (in this case, bullwhip kelp, *Nereocystis luetkeana*) showing the small root-like structure, the holdfast, which it uses to anchor itself to a substrate, for example, a stone or a shell, and the stipe (or stem) that connects it to the leaf-like blades, which need light in order for photosynthesis to occur. The species shown above also has an air-filled bladder, which keeps it upright in the water so that sufficient light reaches the blades.

method with an enormous match-making problem, as the egg and sperm cells need to find each other in water that is often turbulent.

Seaweeds are marine algae

Some species solve the match-making problem by equipping the reproductive cells with light-sensitive eyespots or with flagella so that they can swim. Others make use of chemical substances, known as pheromones or sex attractants. These are secreted and released by egg cells and serve to attract the sperm. Some species, for example, the large seaweed masses in the Sargasso Sea, secrete enormous quantities of slime, which ensures that the egg and sperm cells stick close to each other and do not go astray.

The life cycle patterns of the different species vary greatly and some are complex. That of laver (*Porphyra*), a genus of red alga that grows in the intertidal wash zone of most of the temperate oceans of the world, is among the more intricate.

It is worth recounting a fascinating detail about the life cycle of *Porphyra* because an interesting story is related to its discovery. This discovery was the foundation for an exceptionally successful example of aquaculture that has had enormous nutritional and economic significance and is now worth billions of dollars annually. It relates directly to the cultivation of *Porphyra* for the production of *nori*, which is especially widely used in Japanese cuisine, for example, in the preparation of sushi.

▲ Alga researcher Kathleen Mary Drew-Baker (1901–57) at the microscope.

The blades used in *nori* production grow while the seaweed is in the generation that reproduces sexually, although the organism itself can actually develop asexually from spores. The blades produce egg cells and sperm cells. The egg cells remain on the blades, where they are fertilized by the sperm cells. The fertilized eggs can then form a new type of spores, which are released. These spores germinate into a calcium-boring filament stage that can grow in the shells of dead bivalves, such as oysters and clams, in the process developing spots that give the organism a pinkish sheen. Until 1949 it was thought that this special stage was actually an entirely separate species of alga, *Conchocelis rosea*. Without an understanding of this life stage it was not possible to grow *Porphyra* effectively in aquaculture.

It was an English alga researcher, Dr. Kathleen Mary Drew-Baker (1901-57), who discovered this particular segment of the *Porphyra* life cycle. It is, therefore, not without reason that in Japan she is known as 'The Mother of the Sea'. Even after all these decades, the Japanese *nori* fishers have not forgotten her. Every year, on the 14th of April, they gather on a hill overlooking Ariake Bay in southern Japan to honor Drew-Baker, whose work was instrumental in ensuring the survival of their industry.

The biology of algae

◂ Every year, on the 14th of April, the Japanese pay homage to 'The Mother of the Sea', the English alga researcher, Kathleen Mary Drew-Baker, who identified the underlying biology that made possible the cultivation of *Porphyra* in a predictable manner for the production of *nori*.

PHOTOSYNTHESIS AND SEAWEED GROWTH

Photosynthesis enables seaweeds to convert sunlight into chemical energy, which is then bound by the formation of the sugar glucose. Glucose is the building block for the seaweeds' carbohydrates and, at the same time, an energy source for the production of other organic substances that the seaweeds need in order to grow and to carry out life processes. The photosynthetic process uses up carbon dioxide, which is thereby removed from the water. In addition, phosphorous, a variety of minerals, and especially nitrogen are required. Oxygen is formed as a byproduct, dissolved in the water, and then released into the atmosphere. This byproduct is of fundamental importance for those organisms that must, like humans, have oxygen to be able to breathe.

During the night, when the light level is low, photosynthesis stops and the seaweeds begin to take in oxygen, burn glucose, and give off carbon dioxide. Photosynthesis can still, to a certain extent, be carried out when seaweeds are exposed to air and partially dehydrated. Under normal conditions, photosynthesis is the dominant process, allowing the seaweeds to build up their carbohydrate content. To the extent that they have access to light in the water, seaweeds utilize sunlight more efficiently than terrestrial plants.

The red macroalgae normally grow at the greatest depths, typically as far as 30 meters down, the green macroalgae thrive in shallow water, and the brown algae in between. This distribution of species according to the depth of the water is, however, somewhat imprecise; a given species can be found at a location where there are optimal conditions with respect to substrate, nutritional elements, temperature, and light.

Seaweeds are marine algae

In exceptionally clear water, one can find seaweeds growing as far as 250 meters below the surface of the sea. It is said that the record is held by a calcareous red alga that was found at a depth of 268 meters, where only 0.0005% of the sunlight penetrates. Even though it may appear to be pitch-dark, there is still sufficient light to allow the alga to photosynthesize. In turbid waters, seaweeds grow only in the top, well-lit layers of water, if at all.

Formerly it was thought that seaweed species had adapted to their habitat by having pigments that were sensitive to the different wavelengths of the light spectrum. In this way they could take advantage of precisely that part of the spectrum that penetrated to the depths at which they lived. For example, the blue and violet wavelengths reach greater depths. The red algae that live in these waters must contain pigments that absorb blue and violet light and, as a consequence, appear to have the complementary color red. Experiments have since shown that this otherwise elegant relationship does not always hold true. Seaweed species that live at the ocean's surface may also contain pigments that protect them from the sun's ultraviolet light.

Given that all the substances that seaweeds need in order to survive are dissolved in the water, macroalgae, unlike plants, have no need of roots, stems, or real leaves. Nutrients and gases are exchanged directly across the surface of the seaweed by diffusion and active transport. In some species there is no meaningful differentiation and each cell draws its supply of nutrients from the surrounding water. On the other hand, specialized cell types and tissues that assist in the distribution of nutrition within the organism can be found in a number of brown macroalgae.

Access to nitrogen is an important limiting factor in seaweed growth, particularly for green algae. The increasing runoff into the oceans of fertilizer-related nitrogen from fields and streams has created favorable conditions for the growth of algae, especially during the summer when it is warm and the days are long.

Different species of seaweeds avail themselves of a variety of strategies in order to grow. In some, for example, sea lettuce (*Ulva lactuca*), the cells all undergo division more or less randomly throughout the organism. Others, among them several types of brown algae, have a growth zone at the end of the

◄ Clearing away sea lettuce that would interfere with the 2008 Olympic sailing competitions along the coast of Qingdao, China. As a consequence of a plentiful supply of nutrients and the heat, the seaweeds grew voluminously in a very short period of time.

stipe and at the bottom of the blade; this is where an existing blade grows and new blades are formed. The oldest blades are outermost, eventually wearing down and falling off as the seaweed ages. As a result, the stipe can be several years old, while the blades are annuals. This growth mechanism also allows the seaweed to protect itself from becoming overgrown by smaller algae, the so-called epiphytes, which fasten on to it. On certain species, the epiphytes are found overwhelmingly on the stipes, which can become covered with them, while the blades retain a smooth surface as long as they are young and still growing. Finally, some types of seaweeds, such as bladder wrack (*Fucus vesiculosus*) and the majority of the red algae, grow at the extremities of the blades.

The overall effect of seaweeds on the global ecosystem is enormous. It is estimated that all algae, including the phytoplankton, are jointly responsible for producing 90% of the oxygen in the atmosphere and up to 80% of the organic matter on Earth. We can compare their output with that of plants by looking at the amount of organic carbon generated per square meter on an annual basis. Macroalgae can produce between 2 and 14 kilograms, whereas terrestrial plants, such as trees and grasses in temperate climates, and microalgae can generate only about 1 kilogram.

The productive capacity of macroalgae can possibly be best illustrated by the fact that the largest brown algae can grow up to half a meter a day. That amounts to a couple of centimeters an hour!

Living seaweeds in a Japanese museum

A seaweed researcher working at the Coastal Branch of the Natural History Museum and Institute, a small establishment in the town of Katsuura in the Chiba Prefecture south of Tokyo, has participated in the discovery of a new species of red alga. Even though this is hardly the sort of thing that happens every day, Dr. Norio Kikuchi modestly mentioned his finding only toward the end of my visit, as we were making a tour of his laboratory's temperature-controlled seaweed cultivation chambers and his herbarium.

This new species was found near Osaka in 2003 in six-meter deep water. It has only recently been characterized genetically as a distinct species and, for a time, there was even a discussion about whether it defined a new genus. In 2010 it was, however, decided to classify it as a species belonging to the *Porphyra* family and it was named *Porphyra migitae*. In Japanese, Norio refers to the new species by the poetic descriptor, *akanegumo amanori*, which likens the beautiful, dark red blades of the alga to fluffy cumulus clouds in the evening sky, tinted red by the setting sun.

Norio is an expert in the biological classification of macroalgae and he cultivates a variety of specimens in his small laboratory at the museum. It is in recognition of his efforts that the various species of seaweeds from Chiba figure prominently in the displays of the museum's collections. Seaweeds, whether red, green, or brown, are a daily presence in Japanese life, because they are served at virtually every meal. Norio believes it is an important part of the museum's mission to educate the population, especially children, about the wonderful world of seaweeds. The priority accorded to the subject by this small, modern museum is in striking contrast to their absence at the Natural History Museum in London, the largest museum of its kind in the world. At the latter, seaweeds are not represented in any form in the exhibits, even though a vast, important collection of more than 600,000 specimens can be found behind the closed doors of the Herbarium. Marine algae at present play no major role in the Europeans' outlook on their external environment.

The museum in Katsuura is located right by a small, sheltered bay, on the west coast of the Boso Peninsula, facing the Pacific. At low tide a wonderful rocky shore covered by red, green, and brown algae is exposed.

▲ A newly discovered red alga that was recently named *Porphyra migitae*.

Here children and adults can encounter, in a natural setting, some of the seaweeds, such as the brown *hijiki* and the green *ao-nori* (*Ulva* and *Monostroma*), that form part of their daily diet.

On the opposite side of the Boso Peninsula, facing toward the east and Tokyo Bay, there is another museum at a small place called Hitomi. As the museum stands on a spot that used to be under water, one might say that the seaweeds literally moved into the museum. Until 1961, the area was a sandbank where a local fishing community raised *Porphyra* from which they made *nori*, as they had done for centuries. But the sandbank had to give way to a major land reclamation project to meet the needs of the industrialization taking place in post-war Japan.

The history of the earlier fishing community is now told in this lovely little *nori* museum, which is housed in a dreary, anonymous concrete building. I went to see it on a Sunday morning and, much to my surprise, my faithful guide Norio, my wife, and I were the only visitors. The museum is actually quite fascinating, especially for a seaweed enthusiast, but it became even more so when the office assistant, who tends the museum by herself, found out that we were interested in the history of *nori* cultivation. She telephoned Sumio Ishii, who showed up at the museum within ten minutes, just as she had finished pouring a cup of coffee for us in the museum's little office—she could hardly have known that I would have preferred Japanese green tea. Sumio arrived, still in his work clothes and with coal black hands, carrying a package wrapped in brown paper under his arm. As he was eager for us to see it right away, he asked the office assistant to unwrap it while he talked away and washed his dirty hands.

Until the harvesting of *nori* was shut down in 1961, Sumio had plied his trade here as a *nori* fisher. Since then, he has worked in one of the new factories that was built on the old sandbank, but he has written a book on the history of the Hitomi fisheries. He proudly showed us his wonderful book, filled with old photographs and drawings, which he himself had produced, and then he started to tell his story. His narrative about the *nori* fisheries of the Hitomi area, translated by Norio, brought to life the history on display in the collections.

The Japanese have harvested *Porphyra* in the wild for the production of *nori* since the 6th Century and cultivation of it, albeit by primitive means, started in the 17th Century. For many centuries, the total output was quite modest and of highly variable quality. Everything was done manually, from the cultivating, harvesting, and preparation of the seaweeds to the production of the *nori* sheets. Three determining developments were to change this, leading to modern *nori* cultivation and the accompanying industrialized production of the finished sheets. The first took place in 1821 at the end of the Edo Period, when Jinbei Ohmiya invented what is known as the pole method. Until then, the *nori* fishers had gathered *Porphyra* at low tide where they could find it, for example, on branches that had become stuck in the sea bottom. The pole method was simple and used split bamboo poles or bamboo branches with twigs. The poles were placed in the seabed on shallow sandbanks so that the twigs were partially above water at ebb tide. Success depended on whether the free-floating spores and *Porphyra* blades happened to attach themselves and continue to grow on the poles and twigs. The fully grown seaweeds could then be harvested from boats or by fishers walking in the water on ingeniously designed stilts that were almost a meter in height.

The second determining event is attributable to Bujiro Hirano who, in 1878, introduced a sort of transplantation technique. He had observed that there could be an advantage in moving the poles with germinating seaweed to a growing area in another part of the sandbank. His method was purely empirical, but it made good sense when seen in the light of the later discovery that *Porphyra* has a life cycle with two very different stages that do not place the same demands on their environment.

The pole method is a vertical technique, but starting in the 1930's this gave way to a horizontal technique using nets that are suspended between poles placed in the water two to four meters apart. The spores settle on the nets and grow on them. The outstretched nets cover a much larger surface area of the sea than the poles, giving the *Porphyra* better access to sunlight and nutrients. The production of *nori* increased tenfold over a short period of time after the horizonthal net method was adopted.

Nevertheless, *Porphyra* cultivation remained a low-technology enterprise relying on empirical methods until about the 1960's. At this point mechanical harvesting was introduced and cultivation took full advantage of Kathleen Mary Drew-Baker's scientific exposition of the

▲ Jinbei Ohmiya who invented the pole method for the cultivation of *Porphyra* for *nori*.

▶ ▶ Pictures from the *nori* museum in the Japanese city of Hitomi.

30

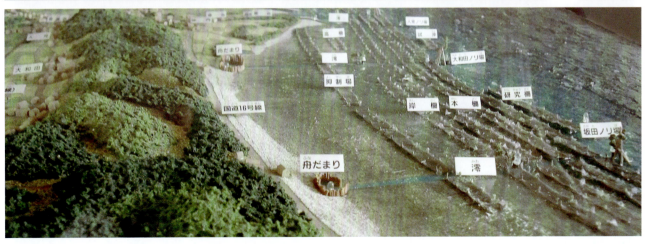

Porphyra life cycle. Her work was the third and most important determining factor in the development of the *nori* industry. This aspect of *nori* history is also presented at the museum in Hitomi.

Until Drew-Baker's discovery came to be known and exploited, there were years when the *nori* crop failed completely and no one could understand why. A poor year was disastrous for the small communities of seaweed fishers. In particular, 1948 was a catastrophic year for the fishers in the Ariake Bay in southern Japan. Chemical pollutants from agriculture ruined their pearl oysters, typhoons smashed their boats to bits, and the *Porphyra* did not grow. Many of them had to abandon their livelihood and seek work in the coal mines.

Drew-Baker's identification of the *Conchocelis* stage made it possible by rational, artificial means to effect the germination of the conchospores, at first outdoors and later indoors. The biotechnological principles involved are described in detail later in this chapter. Using these procedures, the *nori* fishers are generally able to place two successive sets of nets in the same location during a growing season.

Even though the cultivation of *Porphyra* and the history of the *nori* fisheries have become the stuff of museums, a surprisingly large number of the original processes and manual methods have been incorporated as elements of the high-tech cultivation and mechanized harvesting of the raw product and assembly-line production of *nori* sheets. A small, modern *nori* factory can turn out 10,000 sheets an hour, while a traditional manual worker could only make 100 sheets an hour. But both the machine and the laborer use the same type of bamboo mats and the size of the *nori* sheets is as it has always been. Drying now takes place quickly with the help of a stream of warm air, whereas it used to be done in the open air on racks turned toward the sun. The appearance of the finished product is very circumscribed by tradition. *Nori* sheets have a smooth topside and a more matte underside that clearly shows the imprint of the bamboo mat.

Everywhere in the world, Japanese food is served with *nori* in one form or another. But only a few people are aware that in Japan over 10 billion sheets of *nori* are produced every year, the product of an industry steeped in tradition and founded on the aquaculture of a red alga with a remarkable biological make-up.

▲ *Nori* sheets for sushi.

32

▶ Seaweeds surviving under extreme conditions in a 'blowhole' on the west coast of South Island, New Zealand.

ROBUST MARINE SURVIVORS

Seaweeds are found in oceans all over the world. Every species has its preferred habitat, but some of them can get by in almost any location. They thrive especially well in nutrient-rich, cold waters. Interestingly, there are more species of seaweeds in the northern Pacific Ocean than in the northern Atlantic. It is assumed that this disparity is due to two factors: the Pacific is an older ocean and it is less affected by ice formation than the Atlantic. The difference is likely to be evened out over time as a result of natural dispersal and the incidental transfer of species from one ocean to another as a byproduct of dense shipping traffic.

As the majority of seaweeds grow where the land meets the sea, they are constantly exposed to dramatic conditions such as dehydration, physical wear and tear, and drastic changes in temperature and salt content. In order to survive under these circumstances, the cells of seaweeds are equipped with cell walls that are reinforced by special polysaccharides such as cellulose.

Some species of seaweeds tolerate dehydration better than others. Among them are species that can survive losing 90% of their water content, but are able to reabsorb water and restart the process of photosynthesis only a few hours after they are back in the water. Often several different species of seaweeds grow in the same location so that when the tide goes out the more robust types can form a protective and hydrating layer on top of the more delicate varieties. Some brown algae have moisture-absorbent polysaccharides that help to retain water and they can excrete a thick layer of slime on their surfaces to keep them moist and protect them against attack by microorganisms.

▶▶ Gotts Island, a tiny speck of land on the coast of Maine, USA.

34

Seaweeds can also endure major variations in temperature. Typically, the difference can be 20–30°C over a 24-hour period and even more in the course of the seasons. Some are so robust that they can survive freezing temperatures. In order to do so, the seaweeds take advantage of a series of substances, for example, polysaccharides, that function like anti-freeze. Under less favorable conditions, certain seaweeds can change their reproductive strategy in order to ensure that they produce more adaptable offspring.

Seaweeds are marine algae

Tidal variations can be very significant, ranging from less than one meter in some parts of the world right up to 17 meters in others. These variations, together with ocean currents, storms, surf, and being thrown against cliffs, rocks, and sand by the waves, create conditions that severely test the physical strength of the organisms, as well as their ability to remain anchored in place. Some species have adapted to these trying living conditions by developing a structure that is soft and tough, but very flexible. An example of this is the stipe of the brown alga bullwhip kelp (*Nereocystis luetkeana)*, which can stretch by up to 30% of its length without breaking. On the other hand, the turbulent movement of the water is very advantageous for the seaweeds because the currents stir up and renew their nutrient supply.

Marine algae have a noteworthy ability to put up with changes in the salt content of the water. As with temperature, there can be vast daily and seasonal variations. For example, seaweeds that grow in depressions in a rocky coastline can become isolated in small water holes either by tidal or seasonal flux in the water level. In these water holes, the salt concentration goes up or down in the course of the day as water evaporates or when there is a sudden downpour. Some seaweeds grow in the saltiest seas such as the Mediterranean with a salinity of 4.5%, others in the great oceans with one of ca. 3.5%, and still others in less salty waters such as those of the Baltic Sea and some bays, which have a salinity of only 0.3–1%. As long as the seaweeds have an adequate supply of nutrient salts and can obtain energy from photosynthesis, they are able to deal with some variations in salt concentrations by regulating the content of substances such as ions, sugars, and amino acids in their cells. In this way, the changes in osmotic pressure are evened out. Most species prefer high salinity and only a few types can survive in very salt-deficient waters.

Most seaweed species are annuals, living for just a single season or even less. *Porphyra* and *Ulva*, for example, last for only a few weeks or months. Nevertheless, many species are perennials, surviving for from 5 to 30 years, or even longer. In the course of their lifetime, some parts of these seaweeds wither or are shed, while the organism renews itself from the bottom up. Even though bullwhip kelp (*Nereocystis luetkeana*) is an annual, it grows so quickly that it can attain a length of 30 meters and a whole kelp forest can shoot up in the course of a summer.

Seaweeds have developed a variety of defense mechanisms against grazing fish and shellfish. Some have very rapid reproductive mechanisms that allow them to reproduce and disperse their offspring before they can be eaten. In addition, their tissue is often tough and in some cases calcified, so there is little temptation to take a bite out of them. Marine algae commonly situate their most tender parts in the top layer of the water, to which few grazing animals, such as snails, have access. Finally, some species have developed chemical defense mechanisms and contain or excrete special substances that protect them from attack by microorganisms and herbivores. An example of this is the sulphuric acid released by the brown alga stringy acid kelp (*Desmarestia viridis*), which seems to help repel sea urchins.

SEAWEEDS AS ECOLOGICAL DYNAMOS

Collectively, seaweeds constitute an almost inexhaustible pantry for many other forms of life. Large and small creatures, such as krill, bivalves, sponges, and whalesharks, all filter feed bits of marine algae from the water, while grazing animals, such as snails, sea urchins, and herbivorous fish, like carp and tilapia, eat seaweeds directly. The herbivores may, in turn, be preyed on by carnivores, moving the original organic substances produced by the algae higher up in the food chain. Seaweeds also secrete slime into the water. The slime, which consists of polysaccharides together with sugars and amino acids, is ingested by sponges, bacteria, and other unicellular aquatic organisms.

Large kelp forests help to create safe, calm hydrologic conditions and serve as nursery sites and hiding places for fish fry and shellfish. The shed organic deposits associated with these dense growths form sizeable deposits that are of significant nutritional value for other organisms. Snails, worms, and bivalves act as garbage collectors, converting the detritus from plants, algae, and other animals into food. The ideal of sustainable aquaculture is based on incorporating a variety of marine species into this type of nurturing environment to achieve an optimally balanced ecosystem.

Seaweeds can be put to use as a means of ameliorating the environment. As they grow, seaweeds are able to absorb and chemically bind substances like the nitrates and phosphates sometimes found in excess amounts in the run-off from fields and commercial fish farms. Some researchers anticipate that seaweeds could take up a major portion of the excess nutrients associated with intensive fish farming, resulting in their much more rapid growth in such a habitat. In addition, seaweeds can act as decontaminants, by adsorbing a range of environmental toxins, for instance, heavy metals.

Seaweeds are marine algae

Seaweeds in the wild and in aquaculture

EDIBLE SEAWEEDS AND POISONOUS ALGAE

There are only a few species of algae that are not edible, but some of these are actually very poisonous for mammals. In particular, certain blue-green microalgae are poisonous because they may produce toxins that attack the liver and the nervous system or cause skin irritations. Many of these poisonous substances are very stable and exceptionally difficult to break down. Consequently, they can accumulate in filter-feeding shellfish, which may lead to shellfish poisoning. In contrast to microalgae, virtually all macroalgae are non-toxic. One exception is the brown alga *Desmarestia*, several species of which release so much sulphuric acid that it breaks down other algae and serves as a defense mechanism that keeps sea urchins from eating them.

The best known edible blue-green microalga goes by the trade name Spirulina, which covers a number of species of the genus *Arthrospira*. These blue-green microalgae are not, strictly speaking, algae, but are actually classified as cyanobacteria because they lack a well defined cell nucleus. Spirulina algae have a high protein content that may comprise more than half of their dry weight and is composed of important amino acids. Dried Spirulina, which is now mostly used as a food additive, has been eaten by the inhabitants around Lake Texcoco in Mexico and Lake Chad in Africa as far back as the 1300's. The Aztecs used Spirulina to make a type of dry cake.

Another genus of edible microalga is made up of several species of *Chlorella* which are not cyanobacteria but plant-like, single-celled eukaryotes. They are almost as protein-rich as Spirulina and are valued for their particularly large omega-3 content. *Chlorella* is often used as a food supplement; it is claimed that it has a number of healing properties. In the years around 1950, it was thought that farmed *Chlorella* would become the solution to the world's increasing demands for food, but mass production turned out to be too inefficient and costly.

Edible species are to be found among all types of macroalgae, whether green, red, or brown, and several hundred varieties are currently harvested

▸ ▸ Dried sample of winged kelp (*Alaria esculenta*) from the collections of the Natural History Museum in London.

38

Alaria esculenta

Alaria esculenta (L.) Grev.

Dancing algae—
the story of Hana-Tsunomata

Only a few companies outside of Japan have been able to meet the notoriously demanding Japanese standards and quality requirements for seaweed products, as they go far beyond mere flavor and nutritional value. Color, shape, cleanliness, freshness, mouthfeel, as well as general aesthetic appearance are at least as important. One company that has risen to the challenge is the Canadian company Acadian Seaplants, located on the Atlantic coast in Nova Scotia. Working in collaboration with Japanese experts, it has been able to evolve a line of cultured seaweeds that fully satisfy all the different requirements. Since the mid-1990s, the company has had a successful product on the Japanese market.

First, Acadian Seaplants undertook extensive research to determine how it could take advantage of the pristine waters of the North Atlantic, while at the same time achieving consistency. The solution was to develop land-based and tightly controlled cultivation facilities to farm the algae. After screening hundreds of seaweeds, Acadian Seaplants selected a special wild strain of *Chondrus crispus* (carrageen) for its shape, texture, and color. It is farmed in a tank facility, covering an area corresponding to twenty football fields, that is fed by supplies of clean, fresh ocean water from the organism's natural habitat. The result is a fully traceable product that meets the highest sanitation and food safety standards.

▶ Production facilities in Nova Scotia for *Hana-Tsunomata*, cultured carrageen.

The popularity of this wild strain of *Chondrus crispus* may lie in its appeal to both the eye and the palate. It has a distinct crunchy texture and a milder taste than other sea vegetables such as *wakame*. Its shape and three completely natural colors—pink, green, and yellow—are reflected in its somewhat poetic name, *Hana-Tsunomata*, meaning 'flower *Chondrus*' in Japanese. When fully hydrated it is very decorative, whether presented on its own or when lending an elegant touch to other dishes such as a salad. One might say that this farmed Canadian seaweed exhibits all the desirable characteristics of a prototypical Japanese seaweed salad, *kaiso*.

Similar to the changing color of leaves in autumn, the colors of *Hana-Tsunomata*, *aka* (pink), *midori* (green), and *kiku* (yellow), are all naturally produced by the seaweed. With over twenty years of experience working with this unique *Chondrus*, Acadian Seaplants ensures that each seaweed frond is produced to meet exact color specifications.

Recently introduced by Acadian Seaplants, and sister to the *Hana-Tsunomata* line of products, is *Emi-Tsunomata*, or 'smiling *Chondrus*'. With a similar texture and shape to *Hana-Tsunomata*, *Emi-Tsunomata* is presented in its original pigment form—russet. This natural *Chondrus* offers an abundance of nutritional and functional food qualities including antioxidants, dietary fiber, vitamins, and amino acids.

◄ *Hana-Tsunomota* and *Emi-Tsunomota*—cultivated *Chondrus* in four different natural colors.

▲ Rehydrating a dry piece of *Hana-Tsunomata* in sparkling wine causes the crumpled frond to unfold into its beautiful branched flower shape while it dances on the bubbles. *Hana-Tsunomata* is a special *Chondrus crispus* that comes in three different natural colors. During rehydration in room-temperature water the fronds swell up to six times their dry size and seven to eight times their dry weight.

for human consumption. Seaweeds are a major food source in Asia, especially in China, Japan, and Korea. In Japan, they constitute up to 10% of the population's total nutritional intake. Japanese cuisine avails itself of a broad range of seaweeds, with *wakame, nori, konbu,* and *hijiki* being predominant. Seaweeds are eaten as snacks and in salads, used to make sushi rolls (*maki-zushi*), and cooked in soups.

Other seaweed species are utilized because they contain special substances that can be used in food processing, cosmetics, and medications. These end uses will be discussed in a later section of the book.

Originally, seaweeds intended for human consumption were collected along the seashore or picked in the sea. Those that were eaten fresh were harvested locally and consumed in short order. As seaweeds can be dried and, in that form, kept for a long time and transported easily, they were recognized early on as a valuable foodstuff and became a trading commodity. Over time, the demand for seaweeds, for a multiplicity of purposes, grew so great that for many centuries they have been actively cultivated, especially in the Far East. Nevertheless, seaweed aquaculture has until recently been of modest scope and confined to a few species. But this situation is now changing dramatically and the domestication of marine algae is growing at a rapid rate globally.

It is thought provoking that in terms of overall quantity the cultivation of seaweeds is currently more extensive than any other single form of aquaculture, including fish farming. China is the biggest producer of *konbu* (*Saccharina japonica*), with a yearly output of about 2.5 billion tons. But in economic terms, this is overtaken by the Japanese aquaculture of *nori* (*Porphyra yezoensis*), which is worth about US$ 2 billion annually.

While approximately 13 million tons of wet seaweeds are harvested each year in about 40 different countries around the world, 95% of the total yield comes from just ten of them: China, North and South Korea, Japan, the Philippines, Chile, Norway, Indonesia, the United States, and India. About 80% of the seaweed production originates in Asia, with the balance coming primarily from Europe and North and South America. Species of brown algae predominate (7.4 million tons), followed by red algae (5 million tons), and green algae (20,000 tons). Almost 80% of the seaweed harvest is destined for human consumption and the rest is processed industrially, e.g., for use in the biotechnology sector. Cautious estimates put the number of products based on seaweeds and substances derived from them in the range of several thousand, with a yearly global turnover of between US$ 5 and 7 billion.

▸▸ Cultivation of *Por-
phyra* for *nori* in Ariake
Bay in southern Japan.

Marine algae in integrated aquaculture

To date, the most promising form of aquaculture is integrated multi-trophic aquaculture (IMTA), a system in which several species from different levels in a food chain or a food web are placed in close proximity to each other. The idea is to mimic naturally occurring ecosystems to create an optimal balance between the various elements in it. An example of such a system could consist of an organism that is raised on fodder and, in turn, gives off both organic and inorganic waste products that provide nutrients for, or are extracted by, complementary species.

Seaweeds are marine algae

Experimental IMTA projects combining farmed fish, shellfish, and seaweeds have shown that such systems can increase the total output by up to 50% over that obtained from the monoculture of each. In addition, co-cultures have the potential to mitigate the conditions that lead to eutrophication and harmful algal blooms. This is an exciting prospect, given that fish farms are currently the fastest growing sector of global food production, with the yield increasing by just under 10% annually. Comparable figures for the catch of wild fish and the production of meat are 1.4% and 2.8%, respectively.

The domestication of marine species

Given that humans began to domesticate plants and animals for agriculture, market gardening, and animal husbandry about 11,000 years ago, we take it for granted that almost all of our food is farmed. About 90% of all the terrestrial species that are currently being raised have been in production for at least 2,000 years. Since the Industrial Revolution, the growth in terrestrial domesticated species has been a paltry 3%.

On the other hand, domestication of marine species is a much more recent development. About 430 marine species, or about 97% of those now being utilized, have been domesticated only since the beginning of the 20th Century. In this short period of time, the increase has been close to exponential and it is far from over. The rate of growth in the domestication of marine plants and algae is, at the moment, one hundred times as great as that for terrestrial plants. And as the biomass in the oceans is larger than that on land, the potential for exploitation of this resource is enormous.

It is noteworthy that, relatively speaking, the domestication of land-based organisms as a percentage of the total number of species is much less extensive than that of their marine counterparts. Of the terrestrial species, only 0.08% of the known plants and 0.0002% of the known animals have been domesticated, whereas the comparable numbers for aquatic species are 0.17% and 13% for plants/algae and animals, respectively.

There are several reasons for this state of affairs. One is that a significant number of terrestrial plants and animals are poisonous, whereas many marine plants and algae are edible. Actually, virtually all species of macroalgae are edible and few are truly poisonous for humans. In addition, seaweeds reproduce quickly and abundantly. Another factor that has come into play, and which is likely to do so in a more major way, is that marine species are much less prone than terrestrial species to developing diseases and pathogens that can be transferred to land-based organisms, such as humans and domestic animals.

Porphyra for *nori*, raised commercially in Asia, has a yield of more than 1.5 million tons, of which one-third is harvested in Japan. From an economic point of view it is the most important crop, but in terms of quantity, *nori* is superseded by *konbu* and *wakame*. *Konbu*, the general Japanese term for seaweed sourced primarily from *Saccharina japonica* (formerly classified as *Laminaria japonica*), on its own accounts for over 40% of the total world production of seaweed.

Cultivation of seaweeds is, unfortunately, not without risk for the Earth's ecological systems, as the commercially grown species can crowd out the wild varieties, thereby reducing biodiversity. In addition, the regular harvesting of both wild and cultivated seaweeds, if not properly managed, can lead to adverse conditions for those animal and plant species that live in ecological balance with the algae.

Seaweeds in the wild and in aquaculture

SEAWEED CULTIVATION IN THE WESTERN WORLD

In earlier times, it was the practice on the northern and western coasts of Ireland to 'plant' stones in the tidal zone to promote the growth of seaweeds. But overall there has been no well-developed tradition of cultivating seaweeds in the West. A principal reason for this is that algae have never been a preferred food source. Also, cultivation is a labor-intensive process that requires special techniques with which there is little experience in the West. Until recently, seaweed cultivation has been a low-tech activity, involving algae placed directly in the sea without first having been propagated in tanks.

Reasonable quantities of brown algae are produced in France, Mexico, and the United States. In Europe there has recently been a successful attempt to cultivate dulse (*Palmaria palmata*) along the coastline of northern Spain; this endeavour is now responsible for the largest production of red algae outside of Asia. There is also a modest harvest of red algae in France and Portugal.

Harvesting seaweeds on the edge of the Pacific

This digression is all about my trip to Bamfield, a tiny community located on Vancouver Island, Canada, at the mouth of Barkley Sound, directly facing the Pacific Ocean. Some people have heard of Bamfield only because it was the place where the longest underwater telegraph cable in the world came ashore after its journey of over 7,000 kilometers across the seabed from Fanning Island in New Zealand. The cable was laid in the period from 1879–92 and was instrumental in connecting Sydney, Australia, to London, England, via New Zealand and Canada. Others know it as the site of a marine biology research center, where a large staff of researchers and students work all year round, studying the sea and the creatures that live in it. Bamfield is, however, best known as one of the premier sports fishing locations on the West Coast of North America, especially for catching salmon and halibut.

But fishing is not what attracted me to Bamfield; instead, I went there to meet Louis Druehl, who was the first person outside of Asia to approach the cultivation of seaweeds in a scientific manner. Getting to Bamfield was a bit of a journey, because it can only be reached from the land side via an 85 kilometer long pot-holed gravel road, which connects it to the nearest urban center, Port Alberni. An adventure that made the whole trip worthwhile awaited me, as the cold waters of this area are home to one of the world's richest profusions of different species of brown algae.

Louis Druehl began to harvest and grow algae in Bamfield in 1981. One might say that he had the perfect background for this enterprise, having been a professor of marine botany at Simon Fraser University, near Vancouver, for 30 years. His area of expertise is brown algae, which he prefers to call protists, with regard to their evolution, genetic make-up, and cultivation. He is very well known in his field, so much so that a species of seaweed has been named after him. Louis has adapted Japanese cultivation techniques to the conditions on the West Coast of Canada and he now operates a small company, Canadian Kelp Resources, which, apart from continuing to experiment with new ways of cultivating different species of brown algae, harvests and processes the local, freely growing

▸▸ Images from Bamfield and Canadian Kelp Resources on Vancouver Island, Canada, where Louis Druehl harvests giant algae—bullwhip kelp and giant kelp.

seaweeds. Although the company's products are sold mostly in health food stores, Louis takes great pride in their having been chosen by Tojo, one of Vancouver's premier Japanese chefs, for use in his restaurant in preference to imports from Japan. The total annual production is manageable, amounting to about 400 kilograms dry seaweeds a year, primarily the local *konbu* (*Laminaria setchellii,* wild North Pacific *konbu*), bullwhip kelp (*Nereocystis luetkeana*), and giant (or macro) kelp (*Macrocystis pyrifera*). In addition, small quantities of winged kelp (*Alaria marginata*) are gathered and then roasted and ground. Louis is also trying, on an experimental basis, to grow some new species containing substances that have potential for exploitation in the fields of nutrition, cosmetics, and pharmaceuticals. He is particularly interested in their omega-3 fatty-acids content.

Together with his wife, Rae Hopkins, Louis runs Canadian Kelp Resources out of a small cluster of wooden buildings, one of which doubles as their home. The enterprise has a breathtakingly beautiful location by the edge of a cliff with a jungle-like growth of vegetation that leads right down to a little fjord.

Louis collected me at about 8:30 in the morning outside of his house to take me out in his old motorboat, which he has used all this time to plant and harvest kelp all around Barkley Sound. We were lucky with the weather. It was only slightly windy and, as we sailed from the dock, the morning sun broke through the overcast sky, dispelling the threat of rain. It was the middle of September and the harvesting season for one of my favorite seaweeds, bullwhip kelp, was still in progress.

Bullwhip kelp is a fast-growing annual type of seaweed. With its characteristic long, whip-like stipe, large gas-filled bladder, and elongated, thin blades, it resembles a big onion with long hair. Bullwhip kelp likes to grow close to the rocky coastal cliffs and thrives in the turbulent, heavy surf where a mass of nutrients are swirled upward from the deepest and coldest layer of water. Louis adroitly steered the boat in toward the kelp forest, pulled some of the large specimens out of the water, and cut off the meter-long blades just above the air bladder. He let the remaining part of the seaweed slip back into the water, so that it could live on and grow new blades. The harvesting was done as gently as possible, in complete harmony with all the best practices regarding sustainability and respect for the environment.

▲ Wooden shed at Canadian Kelp Resources in Bamfield.

In another spot, in calmer waters, we found some giant kelp, which is recognizable by its abundance of air bladders and large, wide ruffled blades. The ruffles create turbulence in the water as it streams past the blades, thereby ensuring them of a better supply of nutrients. This is an example of the many small, marvellous ways in which organisms optimally adapt to their surroundings. Once the harvest was completed, we headed back, along the way passing by sea lions lazing in the warmth on the many tiny rock outcroppings in the heavy surf by the shore. It turned out, though, that the other frequent guests in these parts, gray whales and orcas, were nowhere to be seen.

When we a arrived back at Louis' place, I was given a tour to explain and demonstrate the simple and careful way in which the harvest is handled. The individual pieces of seaweed are not washed, but are immediately placed on outdoor racks to dry in the sun for a day or so. What happens here is not really just a drying process; the ultraviolet rays of the sun help to convert the polyphenols found in the seaweeds to simple tannins, which are an important component of the way they taste. The seaweeds are then hung up in a proper drying room, where they are kept for about a day at a temperature of 40°C, in the process losing almost 90% of their weight. The room itself is filled with a delightful aroma that a seaweed lover finds irresistible—it is reminiscent of a fresh sea breeze with subtle hints of smoke and toast. I could not prevent myself from pinching a few bits that had been left behind on one of the nails on which the algae are suspended.

From the drying room we proceeded directly into another small room located right next to it. Rae works here, readying the different types of algae and seaweed flakes for shipping. The long, desiccated pieces are cut into appropriate lengths with a pair of scissors, packed into plastic bags, and heat sealed. They are then placed in boxes and stowed away in the attic of the little house. This is a prototype cottage industry, using simple procedures that are all carried out manually.

The dried seaweed products are shipped from the tiny post office in Bamfield, a settlement of only a few hundred people, where the postmaster himself licks and affixes the stamps on the packages. I had secured my own supply of Louis' and Rae's algae and would see to its safe journey home myself, in my suitcase.

▲ The small bladders on giant kelp (*Macrocystis pyrifera*).

SEAWEED CULTIVATION IN ASIA

In Asia, manpower is plentiful. This has enabled the people of Asia, over a period of hundreds of years, to acquire a great deal of experience with various methods of growing and harvesting seaweeds, an undertaking that is normally very labor-intensive.

Seaweeds are marine algae

The two seaweed crops with the greatest yield in Asia are the red alga *Porphyra*, which is used in the production of *nori*, and brown algae, especially *konbu* (*Saccharina japonica*) and *wakame* (*Undaria pinnatifida*). The principal sources of red algae are China, Japan, Korea, and the Philippines, while *konbu* is cultivated intensively in China, Japan, and Korea. Green algae are produced only to a limited extent, mainly in Korea, Japan, and the Philippines.

▶ Harvesting of *konbu* (*Saccharina japonica*) on Hokkaido in Japan.

Since the 1950's, the Chinese have cultivated *konbu* using two different methods of rope cultivation. The ropes are either suspended from floating pontoons and held in place with weights or are kept afloat with the help of buoys. Each method has its advantages and its drawbacks. Cultivation on the ropes that hang into the water is advantageous in clear waters with strong currents, but it can lead to uneven growth.

Konbu is the Japanese word for what is generally called kelp in English. Many, however, will probably be more familiar with the transliteration kombu, which has gained wide currency, but is not quite as precise.

There are several farmed variants of *konbu*, all belonging to the alga order Laminariales with *Saccharina japonica* as the most common. After harvest, the *konbu* is sun dried and the best quality is aged in cellars (*kuragakoi*) from one to ten years, typically two years. During ageing, the seaweed matures and obtains a milder flavor and a less strong taste and smell of the sea. Of the many different variants of Japanese *konbu*, *ma-konbu*, *rausu-konbu*, and *rishiri-konbu* are considered to be of the highest quality and, consequently, are also the most expensive.

NORI—AN AQUACULTURE SUCCESS STORY

Porphyra, a red alga known popularly as laver, is the basis for the valuable sea-weed product *nori*. There are about 70 different species of *Porphyra*, although, at the moment, the number is shifting on account of the reclassification of the species. Twenty-eight different types of *Porphyra* are found along the coastline of Japan and, of these, two are now grown commercially, yielding an annual harvest of about 500,000 tons. They are cultivated on nets suspended in the sea, with more than 600 square kilometers of coastal area devoted to this form of aquaculture. The combined total production of *Porphyra* in Japan and China makes this alga the single most valuable marine crop in the world.

This type of efficient aquaculture dates back only to the 1950's, as earlier attempts to establish *nori* culture on an industrial scale had failed due to insufficient knowledge about the complex life cycle of *Porphyra*. It was made possible as a result of the discovery in 1949 by a British researcher, Kathleen Mary Drew-Baker, that it involves a stage in which the algal spores bore into bivalve shells.

Cultivation of *Porphyra* starts in the spring with the collection of spores from the alga. They are suspended in water and sprayed onto the surface of clean shells that are placed in growing tanks where temperature, light level, and salt content are all carefully controlled. At a temperature of about 10–15°C, the spores germinate to form microscopic filamentous branches that bore into calcareous shells; this is called the *Conchocelis* stage. After the seedlings have developed sufficiently in the shells, they are placed on ropes that are suspended on bamboo poles. These are also submerged in tanks that are kept indoors for the next five months, in a temperature-controlled environment that should preferably not go below 23°C. If the conditions are right, this phase results in the release of conchospores. At this point, nets placed on large rotating drums are moved through the water at a speed that is tuned to the rate at which the conchospores can attach themselves to the filaments of the nets.

Before the nets are placed in the sea, the *Porphyra* culture undergoes a preparatory process. The nets are dipped in the water and then air dried to optimize the conditions for germination. When the seedlings are 1–3 centimeters long, the nets are ready for placement. It is also possible to preserve these nets for later use. To do so, the nets are first dried to about 20–40% of the wet weight of the germinated seaweeds. They are then packaged in plastic sacks and frozen to a temperature of -20°C. On account of the low water content, freezing does not damage the seaweed cells because amorphous ice, rather than water crystals, forms around them. The frozen, cultivated nets can be

Seaweeds in the wild and in aquaculture

The red alga *Porphyra* for the production of *nori* has been cultivated in Japan since the 1600's, but until the 20th Century with limited success. Consequently, most *nori* was earlier harvested in the wild and was considered a luxury good. It is said that in the Edo Period (1603–1868) the price of a single, processed sheet of *nori* was equal to that of one and a half kilograms of rice.

kept for placement in the sea at a suitable point later in the *nori* season. This reduces the amount of labor needed for this part of the process, while permitting the *nori* fishers to set out new cultivated nets on an on-going basis to create the conditions for a more stable output. Typically two sets of nets are placed successively in any given growing season.

Seaweeds are marine algae

The seeded nets are then transferred to places along the open coastline and stretched between bamboo poles, steel pipes, or floating pontoons. The nets are affixed in such a way that tidal changes cause the algae to be exposed to the air for a few hours daily. Cold ocean currents and a water temperature between 3 and 8°C are required for optimal growing conditions. The algae grow slowly at first, but after a scant two months' of growth they have reached a size of 10–20 centimeters. The blades are brownish or purple in color and shaped like long, wide ribbons or plump rosettes. Even though they are very thin, they are fairly robust and difficult to tear to pieces when they are fully grown.

Porphyra is harvested in Japan in the winter from November until March. The harvest is carried out from small boats that can sail in under the stretched-out nets. The algae are cut off the strings on the nets and then washed thoroughly in clean salt water. Each day's crop is processed immediately. Among other products are the fine, thin sheets of dried *nori*, produced by a process that resembles paper making. *Nori* is eaten as is or used to make sushi rolls.

The chemical composition of seaweeds

A SYMPHONY OF GOOD THINGS

Seaweeds are made up of a special combination of substances, which are very different from the ones typically found in terrestrial plants and which allow them to play a very distinctive role in human nutrition. The main reason is that their mineral content is ten times as great as that found in plants grown in soil and, as a consequence, people who regularly eat seaweeds seldom suffer from mineral deficiencies. In addition, marine algae are also endowed with a wide range of trace elements and vitamins. Because they contain a large volume of soluble and insoluble dietary fiber, which are either slightly, or else completely, indigestible, seaweeds have a low calorie count.

Marine algae possess a fantastic ability to take up and concentrate certain substances from seawater. For example, the iodine concentration in *konbu* and other types of kelp is up to 100,000 times as great in the cells of the

▸ ▸ *Porphyra* culture in Japan.

seaweeds as in the surrounding water and the potassium concentration is

20–30 times greater. On the other hand, the sodium content is appreciably lower than that of salt water.

Depending on the species, water makes up 70–90% of the weight of fresh seaweeds. The composition of the dry ingredients in the different types of seaweeds can vary a great deal, but the approximate proportions are ca. 45–75% carbohydrates and fiber, 7–35% proteins, less than 5% fats, and a large number of different minerals and vitamins.

Seaweeds are marine algae

Broadly speaking, the proteins contain all the important amino acids, especially the essential ones that cannot be synthesized by our bodies and that we, therefore, have to ingest in our food. *Porphyra* has the greatest protein content (35%) and members of the order Laminariales the lowest (7%).

Three groups of carbohydrates are found in seaweeds: sugars, soluble dietary fiber, and insoluble dietary fiber. Many of these carbohydrates are different from those that make up terrestrial plants and, furthermore, they vary among the red, the green, and the brown species of algae. The sugars, in which we include sugar alcohols such as mannitol in brown algae and sorbitol in red algae, can constitute up to 20% of the seaweeds. The seaweed cells make use of several types of starch-like carbohydrates for internal energy storage; again, these vary according to species. For example, the brown algae contain laminarin, which is of industrial importance as it can be fermented to make alcohol.

In comparison with fruits and vegetables grown on land, the total fiber content found in seaweeds is relatively high. For example, dried dulse contains even more fiber than oat bran. Brussels sprouts, which have a large fiber content, are surpassed in this regard by the brown algae *hijiki* and *wakame*.

Soluble dietary fiber, which is situated in between the seaweed cells and binds them together, constitutes up to 50% of the organism. Composed of three distinct groups of carbohydrates (polysaccharides), namely, agar, carrageenan, and alginate, it can absorb water in the human stomach and intestines and form gelatinous substances that aid in the digestive process. Insoluble dietary fiber derived from the stiff cell walls of the seaweeds is present in lesser quantities, typically amounting to between 2 and 8% of the dry weight. Cellulose is found in all three types of algae and xylan in the red and green ones.

The primary mineral components in seaweeds are iodine, calcium, phosphorous, magnesium, iron, sodium, potassium, and chlorine. Added to these are many important trace elements such as zinc, copper, manganese, selenium, molybdenum, and chromium. The mineral composition, especially, varies significantly from one seaweed species to another. *Konbu* contains more than 100–1,000 times as much iodine as *nori*. On average, dulse is the poorest choice in terms of mineral and vitamin content but, on the other hand, it is far richer in potassium salts than in sodium salts. In general, marine algae are a much better source of iron than foods such as spinach and egg yolks.

An abundance of vitamins is present in seaweeds, namely, vitamins A, B (B_1, B_2, B_3, B_6, B_{12}, and folate), C, and E, but no vitamin D. The amounts vary by species, as well as with the seasons of the year. For example, *nori* and winged kelp contain much more vitamin A (beta-carotene) than dulse and *konbu*, and *nori* is also richer in vitamin C than oranges. In addition, *nori* may be an important source of vitamin B_{12}, which is not found in terrestrial plants and which many people in Western countries obtain solely from eggs, meat, and dairy products. There are indications that the algae themselves cannot synthesize vitamin B_{12}, but obtain it from the bacteria that live on their surface or absorb it from the surrounding water where it is secreted by marine micro-organisms. Some research appears to show that the vitamin B_{12} found in algae is only a chemical analogue of the 'real' vitamin that is important for human nutrition. Even though there are many vitamins in seaweed, it is not normally possible to meet one's recommended daily allowance solely from this source.

In general, seaweeds contain little fat, with the green varieties usually having the least. The different species of seaweeds also have an assortment of pigments that, together with chlorophyll, help to produce their distinctive colors. An example of this is the brownish-yellow carotenoid fucoxanthin found in brown algae.

One of the amino acids found in seaweeds, taurine, deserves a special mention. While it is not abundant in the kinds of foods usually eaten in the West, it is very present in Asian cuisine, especially in fish and shellfish and, of course, in seaweeds, particularly the red algae. Taurine is not one of the building blocks from which proteins are made, but it plays an important metabolic function in the formation of bile salts, which bind with cholesterol molecules. In this way, excess cholesterol is excreted from the body and the cholesterol level in the bloodstream is lowered.

Seaweeds contain a great deal of dietary fiber, much of which is insoluble and of little nutritional value. Nevertheless, it provides significant health benefits by assisting in the passage of food through the stomach and intestines and facilitating the accumulation of minerals. Apart from its food value, soluble seaweed fiber, in particular the carbohydrates agar, carrageenan, and alginate, have many significant industrial and medical applications, which are discussed in the final chapter.

Overall, it can be said that marine algae are an extraordinary source of those substances that, in every respect, have enormous nutritional value and help to promote good health. I will return to this subject in the next chapter.

The chemical composition of seaweeds

Tables listing the nutritional elements found in the different varieties of seaweed are included in the final section of the book.

The group of blue-green microalgae, which go under the trade name Spirulina, are in a nutritional category of their own. Spirulina contains a high proportion of proteins and convertible carbohydrates, often more than 50% and about 25%, respectively, as well as the same minerals and vitamins as are found in seaweeds. Although it has no iodine content, Spirulina is rich in iron. There is a balance between potassium and sodium salts in Spirulina, which also contains significant quantities of the omega-3 and omega-6 essential fatty acids.

Proteins from the ocean

The sheltered Ariake Bay lies on the coast of Kyushu, the southernmost of Japan's main islands. It is protected from the East China Sea by the Shimabara Peninsula, which from time to time is the site of violent volcanic eruptions. Ariake Bay, which is so large that it could almost be classified as a sea, is very shallow near the coast, with a depth of only about five meters at high tide. As the difference between high and low tide is typically six meters, substantial areas of the seabed are exposed two times daily. Several large rivers run into the bay, making its water rich in nutrients and its sand and mud bottom is home to a varied and distinctive animal population. From November to April, about 30 square kilometers of the bay are covered with nets on which the red alga *Porphyra* is cultivated for the production of *nori*.

Porphyra cultivation is characterized by high productivity and a significant degree of sustainability. A square meter devoted to *Porphyra* cultivation can yield 84 grams of protein, while the most protein-rich land crop, soybeans, can produce only 40 grams and meat from animals a mere 5 grams per square meter. The only protein yield that surpasses *Porphyra* is that of the blue-green microalga Spirulina, which, when cultivated, can yield over 2 kilograms per square meter. Ariake Bay, with its nutrient-laden water and the large tidal range that is vital for the growth of *Porphyra*, is, therefore, a veritable protein factory.

About a thousand *nori* fishers work in Ariake Bay, cultivating and harvesting *Porphyra,* and they account for approximately 20% of the Japanese *nori* production. This represents a value on the order of US $ 400 million. The market price fluctuates in accordance with quality; the most expensive sheets of *nori* can cost up to three dollars each and the cheapest about ten cents.

Saga Prefecture is the site of the biggest *nori* cultivation operation in Ariake Bay. Quite by chance, I had the opportunity to see the *nori* fields in Saga on the very last day of the harvesting season at the end of March, as the remaining nets were being brought in. The season was over because the nutrient level in the water was too low and the water temperature too high. The leader of the Sekei National Fisheries Research Institute in Nagasaki, Dr. Yuichi Kotani, had made contact on my behalf with the Saga Prefectural Ariake Fisheries Research and Development Center. This establishment fulfills a dual function as a marine biology station and a laboratory. Research is carried out concerning the cultivation of *Porphyra*, samples are analyzed, and advice is dispensed to the *nori* fishers. The center had kindly put its research boat at our disposal so that, together with Dr. Kotani and the local *Porphyra* expert, Kuno Katsutoshi, we were able to sail out to have a close look at the areas where the crop is being cultivated. Because Ariake Bay is very shallow, most of the nets are suspended from poles, with only a small proportion anchored to floating systems.

The research boat is a speedboat and within a half hour we were 10 to 15 kilometers from the coast. For most of the trip we were moving through a forest of poles, which create an enormous network of fields with nets stretched out between parallel rows of poles. The nets are about 1.5 meters in width by 30 meters in length, with a net aperture of 15 centimeters. Only a few nets were still in place and, because it was high tide, they lay a couple of meters below the surface.

The harvesting is carried out at low tide using a small punt that is manned by two fishers, one in each end. The boat is manoeuvred sideways between the poles and little by little the net is lifted up over the boat, where the loosely suspended *Porphyra* blades are sliced off by a rotating drum that resembles an upside-down lawnmower. The newly shorn net is then lowered back into the water on a running basis and remounted on the poles.

◄ Harvest and cultivation of *Porphyra* in Ariake Bay in southern Japan.

In the middle of the *nori* field there is an observation tower. Despite the choppy sea, the captain managed, with some difficulty, to bring the research boat quite close to the outside ladder of the tower. We hopped onto the ladder and climbed up to the uppermost observation platform.

The wind was blowing hard and Dr. Kotani did his best to translate Kuno Katsutoshi's explanations, given in Japanese, concerning *nori* cultivation in Ariake Bay. The forest of poles stretched as far as the eye could see, covering an enormous area. It is amazing to think that this is the site of the most lucrative aquaculture in the world.

Kuno Katsutoshi told us that the cultivation method currently used by the fishers consists of hanging a layer of ten nets in the water between the poles and simultaneously attaching, at regular intervals, small nylon bags with shells seeded with the *Conchocelis* culture. When the conchospores are released from the bags, they can fasten on to the nets. The seeded net can then be suspended in the location where it is to be cultivated or dried and frozen for later use.

Once back on dry land, we were shown around the center's small exhibit and museum, which describes the flora and fauna of Ariake Bay, as well as the local fisheries and *nori* cultivation. An assortment of distinctive animals have adapted to life in, and on, the muddy seabed in the intertidal zone. The mudskipper, which the Japanese call *mutsugoro*, is particularly intriguing as it can breathe both in the water and on land. It buries itself in the mud and at low tide can almost run or skate over it. The local fishers have developed a special way of catching mudskippers, which are regarded as a delicacy. They kneel on a wide board and can more or less coast over the slippery exposed seabed. The mudskippers are caught with bait on the end of a long line, using the same principles as fly-fishing, and placed in a barrel on the board.

At the conclusion of my visit to Saga, Kuno Katsutoshi gave me some samples of *hoshi-nori* and *yaki-nori* from the current year's harvest. The sheets were perfect—smooth, soft, and without holes. Their taste was superb and the Saga fishers can be justifiably proud of their product. It had been a good *nori* year in Ariake Bay and I was enormously pleased when they gave me a large package with 50 sheets of this year's Saga *hoshi-nori* as a good-bye present.

▲ Catching mudskippers at low tide in Ariake Bay, southern Japan.

CARBOHYDRATE EXTRACTS—ALGINATE, CARRAGEENAN, AND AGAR

Alginate, carrageenan, and agar play the same role in seaweeds as pectins play in plants. They strengthen the cell walls and create structures in the stipes and blades. Like pectin, these three carbohydrates bond well with water and are thus able to create gels. This property is the basis for the versatile way in which they can be used in the manufacture of a large variety of foods, medications, and technical products.

The three substances are all built up from long, chain-like molecules that consist of many individual groups of monosaccharides. It is the relationship between the different types of monosaccharides that distinguishes them from each other. Alginate, carrageenan, and agar come in many varieties, however, and the precise molecular structure of each one depends on the seaweed species from which it is derived.

Alginates are found in brown algae, whereas carrageenans and agars are predominant in red algae.

FAT CONTENT—LESS IS MORE

Apart from the abundance of minerals in seaweeds, it is their low, but very distinctive, fat content that attracts our attention. It can vary quite widely across the species, for example, being on the order of 1–2% in dulse and *konbu* and up to 4–5% in *wakame*. Within a given species, it also depends on the time of year and the place where it grows. Most of the fats in seaweeds are made up of fatty acids with long hydrocarbon chains.

There are two relationships that are particularly interesting, namely, the relationship between the saturated and the mono- and polyunsaturated fats and that between omega-3 and omega-6 fatty acids.

A common trait of all seaweed species is that they contain about twice as much saturated as monounsaturated fat, but the combined total of unsaturated fat is greater than that of saturated fat. The crucial difference is the content of polyunsaturated fats, especially the superunsaturated fatty acids EPA (eicosapentaenoic acid) and AA (arachidonic acid), which are omega-3 and omega-6 fatty acids, respectively. The polyunsaturated fats make up 30–70% of the total fat content, with omega-3 and omega-6 fatty acids accounting for most of it. It is an interesting point of comparison that no plants contain AA.

The most noteworthy aspect of the fat composition of seaweeds is the balance between the essential fatty acids omega-3 and omega-6. Essential fatty acids are those that our body cannot make and, therefore, has to ingest.

The chemical composition of seaweeds

59

In the different species of seaweeds, the proportion of omega-3 to omega-6 typically falls between 0.3 and 1.8, with variations within a particular species, again dependent on where the seaweeds are grown and the time of year. From a nutritional standpoint, this is close to the ideal proportion for a human diet. Some nutritionists cautiously recommend that the figure should be about 0.2, but others think that it should be close to 1. These recommendations should be contrasted with the proportion of 0.05 to 0.1, which is typical of the average Western diet. As a consequence, this diet is far too rich in omega-6 fats and far too poor in omega-3 fats.

Seaweeds are marine algae

Whereas seaweeds contain fair amounts of EPA (eicosapentaenoic acid), the content of the other important omega-3 fatty acid found in the various seaweed species, DHA (docosahexaenoic acid), is often too small to be measured. This is in contrast to fish, which can have large quantities of both EPA and DHA. These two substances generally make up 30% and 20%, respectively, of the content of the fish oil sold as a dietary supplement. Unlike the macroalgae, the microalgae contain significant quantities of DHA in addition to EPA.

The omega-3 fatty acids found in fish and shellfish are not produced by these organisms themselves but obtained via the food chain from algae.

Sterols are a particular type of fat, which seaweeds, like other higher organisms, utilize to strengthen their cell membranes. Two of these sterols, fucosterol and desmosterol, are related to cholesterol. The brown algae have an especially large sterol content, up to ten times as great as that in red algae. Normally, humans cannot make use of these seaweed sterols and less than 5% of the total is absorbed in the intestines. At the same time, however, they act to reduce the amount of cholesterol that is absorbed from other food. Studies have indicated that seaweed sterols help to decrease free and bound cholesterol levels and, in addition, to lower blood pressure.

Iodine—a key element

Seaweeds contain iodine, although the exact quantities vary greatly by species. The iodine content is dependent on where the seaweed grew and how it has been handled after harvest. Furthermore, the iodine is not evenly distributed, being most abundant in the growing parts and least plentiful in the blades. In particular, the brown seaweeds contain large amounts of iodine. It is not known for certain why brown seaweeds contain so much iodine, but this is probably linked to their capacity for rapid growth. Recent studies of the brown seaweed species oarweed (*Laminaria digitata*) discovered very large concentrations of inorganic iodine in the form of iodide (I^-) in the cell walls. Iodide was found to act as the main antioxidant for this tissue. In addition the study showed that the action of iodide was not accompanied by an accumulation of organically bound iodine.

Unlike seaweeds, terrestrial plants seemingly have no need for iodine and it might be purely coincidental that it is found in the roots and leaves of some plants. The low iodine content in plants is often due to its being present in only meagre amounts in some soils. As a consequence, terrestrial plants are a poor source of iodine and this can result in iodine deficiency in vegetarians and vegans.

The history of the discovery of iodine as an element actually begins with seaweeds. Bernard Courtois (1777–1838), a French chemist, was working in his laboratory in 1811, extracting saltpetre from seaweeds for the production of gunpowder for Napoleon's army. He noticed that his chemical experiments with the seaweed ash gave rise to a violet-colored vapor that condensed as crystals on his copper vessels and, unfortunately, caused them to corrode. Courtois convinced first his French, and later his English, fellow chemists that his discovery had important dimensions. Their work then rapidly led to the conclusive identification of the substance that was the source of the vapors. It turned out to be an element and, as the color violet is called *iodes* in Greek, the new element was given the name iodine (and has the atomic symbol of I).

The accidental discovery of iodine is a wonderful example of how research and an open mind on the part of the researcher can lead to results that have a major significance for the economy and for human health.

GASES AND THE SMELL OF SEAWEEDS AND THE SEA

The Roman poet Virgil is credited with the saying that there is nothing more worthless than washed-up seaweed, popularly expressed as 'nihil vilior alga'. He was absolutely right in so far as dead, rotting seaweeds give off a horrible stench. The unpleasant smell of decomposing seaweeds is due to a number of gases that are not dangerous, but are the source of odors that we consider offensive.

One of the chief culprits is a chemical substance, dimethylsulfoniopropionate (DMSP), which is found in red and green algae, where it helps to regulate the osmotic balance of the cell in relation to the surrounding salt water. Some researchers think that DMSP is an important antioxidant, which provides support for the physiological functions of the algae. DMSP accumulates in those animals in the food chain that feed on seaweeds.

Dimethylsulfoniopropionate has no taste and no smell, but dimethyl sulphide (DMS), a volatile gas that is a byproduct of DMSP breakdown, has a characteristic disagreeable odor. It is formed when DMSP is oxidized in the atmosphere or when it is degraded by bacterial action. It can also be released in the course of food preparation when fresh fish and shellfish are heated. When present in small quantities, dimethyl sulphide is the cause of what we often call 'the smell of the sea', but in large quantities it results in the disagreeable smell that is associated with rotten seaweeds and with fish that is no longer fresh.

Seaweeds are marine algae

Dimethyl sulphide is the most abundant gaseous sulphur compound emitted into the Earth's atmosphere as a result of biological processes. When DMS is released into the atmosphere, it, in turn, is oxidized to form particulate aerosol substances. These can cause condensation of water vapor, which brings about cloud formation and thereby affects the weather. So while we may find its odor offensive, many scientists think that the dimethyl sulphide generated from the decomposition of marine algae, especially phytoplankton, has a vital role to play in the regulation of the Earth's climate.

Seaweeds also contain trimethyl-aminoxide, which is odorless but which is quickly reduced to the foul-smelling trimethylamine, familiar from the ammonia-like odor given off by fish that is decomposing. Generally speaking, seaweeds produce more trimethylamine than any fish, with the exception of sharks. This substance is formed on the surface of the blades of the dying seaweed by the action of enzymes released by the bacteria that live on the organism. This conversion process can be observed when dried *nori* is soaked in water and gradually takes on a less pleasant smell.

When brown algae and some types of red algae decay, it can cause the formation of another sulphurous gas, methyl mercaptan. This is the gas that smells like rotten cabbage and that is often added to natural gas in order to alert us to its presence.

Conversely, fresh seaweeds, much like a delightfully aromatic ocean breeze, have a characteristic, agreeable smell. In both cases this is due to substances called bromophenols, which the seaweeds synthesize. They are released into the air and accumulate in ocean-dwelling fish and shellfish through their food intake. As there are no bromophenols in fresh water, fish that live in lakes and streams lack the same pleasant odor and taste as their salt water cousins.

▸ ▸ Eyjólfur Friðgeirsson harvesting dulse in Iceland.

Revival of a proud tradition

Icelanders have an unbroken tradition, which goes back to the time of the sagas and the first Viking settlements, of valuing the seaweeds found in their pristine waters and using them for human consumption. It is believed that this tradition was carried to Iceland by Norwegian settlers who arrived by boat, bringing with them women slaves from Ireland. Seaweeds were boiled with milk or mixed with barley or oats to make bread. It is said that during times of famine almost half of the bread dough consisted of seaweeds, in particular dulse. These traditional uses are now mostly forgotten and have been on the brink of extinction.

About two hundred years ago, the Icelandic market for dry dulse amounted to about 30 metric tons, most of which was produced by a single family-owned farm in the southwestern part of Iceland. Today the total sold amounts to a mere two tons. While there is still a long way to go to regain the magnificent status it enjoyed in the past, the Icelandic tradition of seaweed consumption has witnessed a revival in recent years due to the dedicated efforts of individual harvesters and some small companies. Of the eleven species of seaweeds traditionally eaten in Iceland, dulse remains, by far, the most popular.

Since 2005, Eyjólfur Friðgeirsson and his company, Íslensk hollusta ehf, just outside Reykjavik, have been spearheading the trend and are now selling dried seaweeds and innovative products based on seaweeds, such as marinated dulse, dulse soy sauce, and kelp snacks. Demand for these is growing, both from the home market and for export. The term 'hollusta' in Eyjólfur's company name signals that he markets his products as healthy and whole foods.

Another family-owned company, Seaweed Iceland (Hafnot ehf), is based in Grindavik on the southern coast of Iceland. The owner, Grettir Hreinsson, focuses on high-quality products, made in small batches and sold in designer packaging. His marketing strategy is typically Icelandic— it emphasizes sustainable harvesting in unpolluted waters, seaweeds of the highest purity, and environmentally friendly processing using geothermal energy for moisture-controlled drying at low temperatures (28°C). Sugar kelp from this company is some of the cleanest I have ever seen.

A couple of years ago, I had an opportunity to experience at first hand the way in which seaweeds can complement and add value to other marine activities. I was contacted by Símon Sturleson and Anna Melsteð of Íslensk bláskel, a company based in the small town of Stykkishólmur facing the wide fjord Breiðafirði in Western Iceland. The fjord has shallow waters and accounts for about half of Iceland's intertidal waters, with tidal variations of up to six meters. The bedrock is a perfect substrate for seaweeds, especially the knotted wrack and tangle processed for alginate at the local plant, Thorverk. These waters are also ideal for farming the blue mussels marketed by Íslensk bláskel, even though, surprisingly, there is no tradition of eating these bivalves in Iceland. Símon and Anna asked if there was any way they could make use of the seaweeds that tend to foul their mussel lines in the summer. We discussed the possible uses of sugar kelp for human consumption and the various ways of exploiting this 'by-catch'. The end result is that the company now has a modest output of dried, high-quality sugar kelp, dulse, and winged kelp, some of which ends up on the menus of top restaurants.

The revival of the proud Icelandic tradition of using seaweeds may turn out to be not only a nostalgic experiment but also a way for Iceland to recover from its recent serious economic setback.

◄ Winged kelp (*Alaria esculenta*) that has been blanched.

Seaweeds & human nutrition

Edible marine algae

*Seaweeds &
human nutrition*

NEARLY ALL SEAWEEDS ARE EDIBLE

Both in terms of overall quantity and commercial value, the greater proportion of all the seaweed types that are exploited by humans are eaten or made into food additives. The rest are utilized for medicinal and industrial purposes.

In contrast to terrestrial plants, the vast majority of seaweed species are edible, but not all of them are equally suitable for human consumption because they are either tough or simply not particularly tasty. A few species can be eaten raw if they are completely fresh and collected from areas of clean water. Most, however, need to be processed in some way, usually by drying, cooking, or toasting, in order to make them palatable. This often results in a noticeably improved flavor.

In this book, we will learn more about some of the most commonly eaten species of seaweeds. Most of them are brown and red algae, but we will also discuss a few of the green algae and a couple of the blue-green microalgae. Depending on where you live, these will probably be available only as either frozen or dried products.

With regard to sources of seaweeds, a few words of caution are in order. Because seaweeds grow so abundantly along almost all of the world's coastlines, it is quite easy and tempting to harvest marine algae for oneself. Bear in mind, however, that doing so calls for real expertise in identifying the various species and taking care to do so without harming the ecosystem. In areas where the difference between high and low tide is small, some of the more interesting varieties might be found only in the deeper water. In such cases, it is necessary either to dive for them or to harvest them from a boat using an appropriate tool. Water that looks to be clean and clear may not be entirely free of harmful substances. Harvesting, therefore, also involves detailed knowledge concerning water quality and the pollutants to which the seaweeds might have been exposed. Seaweeds are best harvested early in their growing season where they are less fouled by other organisms.

We are fortunate that the most readily available seaweeds are ones that can be used in a whole range of ways: as salads, in soups, for sushi, in desserts, in bread, as snacks, and in candy, or as herbs and flavor enhancers. A tea-like infusion can be made from seaweed extracts. Some seaweeds are at their best when prepared in one particular way, while others are versatile and can be used for virtually any kind of food and eaten at any meal.

▲ Kelp: Japanese *konbu*
(*Saccharina japonica*).

Edible marine algae

◄ Drying *konbu*
(*Saccharina japonica*) in
Hokkaido, northern Japan.

Kelp and *konbu*—not at all tough

Kelp, known in Japanese as *konbu*, is probably the seaweed that can most easily be thought of as a vegetable from the sea. The term covers about 300 different types of large brown algae belonging to the biological order called Laminariales, which is subdivided into families, many of which are eminently edible. The name Laminariales is evocative of the long, thin blades, which resemble lamellae, found on most of these seaweeds. In common with some other seaweeds, the color of kelp is well suited to its classification as a brown alga. Kelp grows under the surface of the water and can form enormous forests, anchored to the ocean bed and reaching as far as 50 meters up into the water.

A brown alga called wild Atlantic *konbu* (*Saccharina longicruris*) is native to the North Atlantic. Its blades are thinner than those of Japanese *konbu* and it is easier to cook. In the North Atlantic area, the most popular variety is sea-girdles (*Laminaria digitata*), generally used in soups and stews. On Iceland, tangleweed (*Laminaria hyperborea*), winged kelp (*Alaria esculenta*), and the slightly sweet sugar kelp (*Saccharina latissima*) have traditionally been eaten dried and toasted.

Some species of kelp are better suited for human consumption than others. Their texture varies according to type and age, with some having

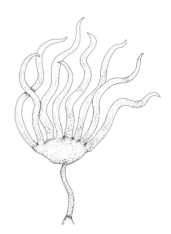

▲ Kelp: sea-girdles
(*Laminaria digitata*).

thin, soft blades, while others are thick and tough. After the kelp has been harvested and rinsed, it is normally dried in the sun, then pressed and packaged whole or in pieces, or ground into coarse granules. When dried seaweeds are soaked in water, their thickness usually swells by 50%.

▲ Kelp: tangleweed
(*Laminaria hyperborea*).

Kelp must be cooked before it is eaten, otherwise it is far too salty and tough. It is generally boiled for 10–20 minutes, which leaves it with an agreeably soft, easy-to-chew texture, without taking away its firmness and shape. While it takes a long time to cook kelp to the point where it starts to dissolve, it does become more bitter the longer it cooks. Kelp contains large quantities of MSG, monosodium glutamate, which imparts the *umami* taste.

Oboro (*tororo*) *konbu* is a specialty product originating in Osaka, Japan. Dried *konbu* blades are marinated in rice vinegar, semi-dried, and then cut into paper-thin shavings with a razor-sharp knife. It can be wrapped around rice or other ingredients and can easily be used as a condiment for a fish dish or simply eaten on its own as a snack.

Kelp is often incorporated into soups, salads, and cooked dishes. In granulated form, it can be sprinkled on food or used as a food additive in many different dishes. It can also be deep-fried in oil to make chips or sautéed and marinated. Kelp lends itself to being cooked with dried beans and other legumes, as it seems to shorten their cooking time and leave them softer and more digestible. Claims have been made that it is the glutamic acid content of kelp that tenderizes the fibers in the beans, but this is doubtful.

Kelp is rich in all the vital minerals, especially calcium, potassium, magnesium, and iron, as well as the trace elements manganese, zinc, chrome, and copper. Because it contains more potassium than sodium, kelp makes a suitable substitute for table salt. In addition, kelp may have a high concentration of iodine and it is, therefore, recommended that one should not eat it to excess. This is especially true of seaweed-based food supplement tablets that may contain very large quantities of iodine. In order to mitigate this problem, kelp is often boiled before it is dried.

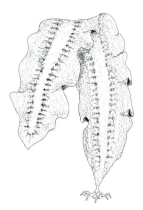

▲ Kelp: sugar kelp
(*Saccharina latissima*).

▸▸ *Oboro* (*tororo*) *konbu* consists
of thin shavings of pale green
konbu (*Saccharina japonica*)
marinated in rice vinegar.

GIANT KELP

A particularly interesting large brown alga called bullwhip kelp (*Nereocystis luetkeana*) is found along the northern Pacific coastlines of the Unites States and Canada. Even though this species is an annual, it is a true giant. It can grow exceptionally quickly, up to half a meter in a day, reaching a length of over 70 meters. The biomass production of a forest of bullwhip kelp is enormous.

The popular name is probably derived from the resemblance of its long flexible stipe with a cluster of thin fronds at the end to the type of whip traditionally used in North America to control livestock. The stipe is anchored to rocks on the seabed and held aloft in the water by a large air-filled bladder, typically the size of a fist, but the largest of which can hold up to three liters of air. About 12% of this air is carbon monoxide. This characteristic bladder of the seaweed is the source of its scientific name, as *Nereocystis* means mermaid's bladder in Greek.

At the top of the bladder, there is a clump of very long, broad, and thin ribbon-like blades. They benefit from the rapid growth pattern of *Nereocystis* and, as a result, are most delicate and eminently edible. When dried and toasted, they are crisp and delicious. The hollow stipe can also be eaten if cut up into rings and toasted or marinated.

Another tasty species of very large brown algae is the perennial giant kelp (*Macrocystis pyrifera*, also known as *Macrocystis integrifolia*) which consists of a very long stipe, to which a series of wide blades are attached by small, pear-shaped, air-filled bladders. The blades are broad and narrow towards the top, with a sawtooth edge and a characteristic wrinkled pattern of waves and bulges. This pattern helps to create turbulence in the water that flows around them, thereby replenishing the supply of fresh nutrients available to the organisms. Dried giant kelp looks like a piece of crepe paper but, amazingly, the original pattern reappears perfectly when the blade is soaked in water.

▲ Bullwhip kelp
(*Nereocystis luetkeana*).

LAVER—THE DELICATE RED ALGA

Because it is used to produce *nori*, laver is probably the seaweed genus that figures most prominently in human nutrition in all parts of the world. Its scientific name is *Porphyra*, which is the Greek word for a purple-red color.

Laver grows in temperate waters in the upper and middle parts of the intertidal zone and is, therefore, easy to harvest at low tide. It has no stipe and consists of ribless blades that fasten themselves directly to rocks. The blades are brown or purplish in color, elastic, and very thin, being made up of only one or two layers of cells.

▲ Giant kelp
(*Macrocystis pyrifera*).

▶ ▶ Underwater forest of bullwhip kelp (*Nereocystis luetkeana*).

Many different species of *Porphyra* are harvested in various parts of the world. They are known by an assortment of popular names, such as *slebhac* in Gaelic Irish, *purpurhinde* in Danish, laver or sloke in English, *zicai* in Mandarin, and *gim* in Korean. In New Zealand, the Maoris have for centuries gathered several species of *Porphyra* that they call *karengo*. In the North

Pacific, *Porphyra perforata* and *Porphyra nereocystis* grow as epiphytes on the much larger *Nereocystis*, while *Porphyra umbilicalis*, also known as purple laver and wild Atlantic *nori*, is found mostly in the North Atlantic.

Porphyra is exceptionally nutritious. Between 30 and 50% of its dry mass is protein, of which about 75% is digestible, and it is rich in vitamins A and C. It contains ten times as much vitamin A as spinach and four times as much vitamin C as apples. The vitamin C, however, degrades quickly in the dried product. *Porphyra* is the variety of edible seaweed that has the lowest iodine content. Depending on the species, it contains more or less equal amounts of omega-3 and omega-6 fats.

Even though the blades are thin, they are usually too tough to be eaten raw. Toasting lightly tenderizes them and brings out slightly sweet and nut-like taste components that are due to alanine, which makes up 25% of the free amino acids in *Porphyra*. Toasted and crumbled *Porphyra* makes a fine addition to soups, breads, and salads.

NORI—A SOUGHT-AFTER *PORPHYRA* PRODUCT

The most widely recognized name associated with red algae of the *Porphyra* genus is the Japanese word *nori*, which is a sought-after food product in Asia and, with the globalization of sushi culture, in the rest of the world as well. *Nori* should not be confused with *ao-nori* (green *nori*) or *hitoegusa*, which is made from green algae (*Monostroma nitidum, Monostroma latissimum, Ulva* spp.) and is used in Japan in the form of small flakes that are sprinkled on warm rice.

In Japan, *Porphyra tenera* and, more recently, *Porphyra yezoensis* are the most important species cultivated to supply its vast *nori* industry. Because of its complicated life cycle, the conditions for *Porphyra* aquaculture have to be followed strictly. During the period when it is harvested, it has blades that are in the shape both of rosettes and of 10–20 centimeter long ribbons.

The algae are washed in fresh water on the day they are gathered, chopped into smaller pieces, and then subjected to a process that resembles the one used to make paper. The chopped pieces are suspended in water to form a thin paste that is poured onto bamboo mats in frames that are ca. 19×21 centimeters in size. This allows the water to drain off and the algal paste forms a thin layer on the mats. In bygone times, there was a very elaborate method for drying the sheets in the sun, but now they are simply subjected to low heat. After this, they can, like pieces of paper, be loosened gently

▲ Laver (*Porphyra* sp.).

In the 1700's, English whalers going on long sea voyages ate dried laver as a way to prevent scurvy. Unfortunately, however, the vitamin C content degrades quickly in dried *Porphyra*.

In Europe, *Porphyra* is harvested primarily in Ireland, Scotland, and Wales where it is a traditional food. The age-old way of preparing it is to boil it and then chop it finely to make a paste called 'laverbread', which resembles puréed spinach. It is used in cooked dishes, mixed into biscuits, and spread on bread.

from the bamboo mats. The finished sheets, called *hoshi-nori* (dried *nori*) or *yaki-nori* (dried and toasted *nori*), are packaged in bundles in an airtight wrapping. A small packet of silicagel is enclosed to ensure that the sheets of *nori* stay crisp and dry. *Nori* is very absorbent and rapidly loses its crispness when exposed to air or dampness. After it has been in contact with water for a few seconds, *nori* becomes soft and quickly dissolves into small pieces. A soft sheet of *nori* can be made crisp again by toasting it for a few seconds in a toaster oven before use.

Nori has a sweet, mildly nutty taste that is only really apparent when it is toasted. It is so rich in vitamins that a single sheet of *nori* has as much vitamin A as three eggs. In addition, *nori* is the seaweed product that contains the most protein.

About ten billion sheets of *nori* are produced annually in Japan. They come in a broad spectrum of qualities, with prices to match. High-quality *nori* sheets are thin, without any holes, of a uniform texture, and have a dark green, glossy surface. Sheets that are thick, matte, full of holes, and reddish in color are of much lower quality. The best Japanese *nori* can easily cost 50 times as much as the cheapest.

Sheets of *nori* are wrapped around cooked rice to make sushi in the form of cylindrical rolls (*maki*-zushi), cone-shaped hand rolls (*temaki*-zushi), battleship sushi (*gunkan*-zushi), or rice balls (*onigiri*). Snacks are also made by folding small *nori* sheets around rice cakes (*senbai*).

A great deal of *Porphyra* is harvested in Japan, Korea, and China to produce *nori*. Japanese *nori* is generally considered the finest and there are three districts that are each famous for their distinctive varieties: Tokyo Bay (Chiba), Kobe Bay (Hyogo), and Ariake Bay (Saga). That from Kobe is usually thicker and has a more pronounced taste. Saga *nori* is very lustrous and moss green with dark spots. The sweet taste and crispness of *nori* from Ariake Bay in southern Japan are attributable to the significant tidal flux in that area, which exposes the seaweeds to the air for many hours each day. These characteristics make it very suitable for use in thin rolls (*hosomaki*-zushi) and hand rolls.

▲ *Nori* produced from *Porphyra*.

Toasted *nori*, which is cut into small pieces or narrow strips, and sometimes mixed with sesame oil or soy sauce, is often sprinkled on salad or rice as a taste enhancer or as a seasoning (*furikake*). Most Japanese meals will be accompanied by *nori* in one form or another.

At sea and in the pub with Japanese nori fishers

For many centuries, the bay at the edge of the Chiba Prefecture has been Tokyo's pantry. Over the years, fishers have been harvesting an abundance of good things from the sea: fish, shellfish, and seaweeds. Edo was the old name of Tokyo and the designation Edomae, Tokyo Bay, has been, and still is, a sign of quality for marine products. Not the least of these is the justly famous *nori* produced from the local *Porphyra*—it is greatly prized by seaweed lovers with discriminating taste buds.

Edomae-*nori* is still of high quality, even though this bay is no longer what it once was and only a small part of its original coastline has been preserved. Japan's rapid industrialization after World War II and the steep economic upturn after the 1960's have left their mark. Manufacturing industries have appropriated much of the coast, and the sandbanks in the bay have been reclaimed as valuable building sites for large factory complexes. The *nori* fishers have been crowded out. In 1965 there were over 9,000 families plying this trade in Chiba, but during the next three decades the number of *nori* fisher families fell by 90% at the same time as the area they cultivated shrank by 75%. This tendency has continued to the present. *Porphyra* is now grown in three small, narrow strips along the eastern shores of Tokyo Bay.

In the middle of March as the *nori* fishing season was drawing to a close, I traveled to the area of the Chiba Prefecture that is near Kisarazu, a city at the mouth of the Obitzu River. Its muddy waters are rich in nutrients, especially nitrates and phosphates, which create the conditions in which *Porphyra* thrives. As is the case in most Japanese *nori* fields, the species that is grown here is *Porphyra yezoensis*. The seaweeds are cultivated on nets suspended between poles implanted in the sandbanks a few hundred meters from the shore. Further out in the deeper water, they are grown on floating nets that are stretched between frames attached to buoys.

I had been in contact with Dr. Norio Kikuchi, a Japanese *Porphyra* researcher, who knows the local *nori* fishers, including one, Norio Kinman, who is a good friend and, coincidentally, has the same first name. To simplify, I will call him Norio-san. He met us one afternoon outside a small,

►► Cultivation and harvesting of *Porphyra* in Tokyo Bay for *nori* production.

76

local *minshuku* called Yohei, a modest, family-run traditional Japanese inn, where we would stay for the night.

We all drove out to the place where Norio-san moors his boats, a few hundred meters from the mouth of the Obitzu River. There were two boats at the jetty. One was equipped for mechanical harvesting of the seaweeds growing on the nets out in the sea. The second one, not surprisingly called *Kinnorimaru* (composed of the Japanese word *kin* for gold, *nori*, and *maru*, a word derived from the now archaic male pronoun *maro*, which is conventionally attached to the name of a commercial boat), is completely open and designed for transport. This was the one in which we set out. As water in the river and covering the sandbanks is very shallow at ebb tide, the boat is flat-bottomed. The propeller of the small outboard motor often dug into the sandbanks and it was necessary to reduce speed. A fresh wind was blowing, so we had put on oilskins and rubber boots for protection. Norio-san steered confidently between the waves and laughed delightedly when a spray of water hit the landlubbers square in the face.

At the mouth of the river, we had a view on one side toward the rows of poles that are characteristic of a *nori* field, placed as they have been for hundreds of years. On the other side, the view is marred by the sight of large, ugly factories and the tall steel towers of an oil refinery. For a Westerner, the paradoxical aspect of Japan is that, in the rush to modernize and chase after economic progress, a great deal of tradition has been thrown overboard and important values relating to Nature and culture have been pushed aside. In his books, *Lost Japan* and *Dogs and Demons*, the author Alex Kerr has given a very illuminating description of this particularly Japanese phenomenon, which he refers to as goverment subsidized destruction and construction driven by an idiosyncratic construction industry.

Norio-san is one of the mere 89 *nori* fishers still left in Kisarazu. At sunrise every morning during the season, typically from November to March, they sail out to their nets. Within an hour or so, each boat has collected and brought back to shore about 600 kilograms of fresh *Porphyra*. The rest of the day is spent in the small local *nori* factory, working the seaweeds to produce sheets of *nori* from the day's harvest. The factory is owned by a fishers' cooperative and its annual output is ca. 50 million sheets of *nori*. This sounds like a significant amount, but the installation was costly and

Nori in Japanese *haiku*

kami atsuru
mi no otoroi ya
nori no suna

failing health
chewing dried seaweed
my teeth grate on sand

Matsuo Bashō (1644–1694)
(translated by Jane Reichhold)

nori prices have stagnated on account of competition from producers in Korea and China. Consequently, *nori* fishing in Chiba is on a downward slope and young men no longer automatically follow in the footsteps of their fathers. Their way of making a living is threatened not only by the industrial invasion of the fertile sandbanks and international competition, but also by the rapid changes to which modern Japanese society is subject.

One of the fishers put it this way in the pub in the evening: my son does not want to be a *nori* fisher—he wants to study. The main problem affecting the industry is lack of manpower. But the fishers of Kisarazu have not given up and have banded together into a non-governmental organization (NGO). This very evening, six of them, led by Norio-san, were meeting at Minshuku Yohei. They were working on a project to teach children about *nori*, how to cultivate *Porphyra*, and what it all means for

◄ Six *nori* fishers at the pub in Kisarazu on Tokyo Bay.

the ecosystems of the bay and the local community's understanding of itself. These points stand out in stark contrast to what we had seen that afternoon of the way in which industrial expansion was inexorably squeezing out seaweed cultivation.

Even though it may seem incomprehensible, neither the *Porphyra* researcher nor the fishers think that there are major pollution problems in Tokyo Bay, especially not in comparison with an earlier time. Wastewater is treated and emissions from the factory chimneys are filtered. The fishers do admit, however, that air-borne dust particles are a concern, because they can be suspended in the water and then settle on the blades of the *Porphyra*. Because it is very difficult to rinse away these particles in their factory, this factor contributes to lowering the quality of the finished *nori*.

Norio-san's NGO group was also working on cultivating *Porphyra tenera*, which yields a more delicate *nori* than *Porphyra yezoensis*. The former is not nearly as robust, however, and consequently it accounts for only a small percentage of Japan's *Porphyra* output. Norio-san's attempts to grow *Porphyra tenera* had not yet been successful, but he was not ready to abandon the experiment.

Nori fishing has lost much of its former romantic aura, but has, on the other hand, become a modern industry where one can earn as much as an urban industrial worker. As the cultivation of *nori* is seasonal work, however, the fishers must, as in earlier times, supplement their income in other ways. For example, Norio-san and his colleagues in Kisarazu also collect mussels in the shallow waters.

Nori fishing is still an occupation in which the same people either are involved in the whole process or keep a close eye on it, from cultivation to growing and gathering, through to the packaging of the finished product. Since the 1960's, harvesting has been done with the help of machinery and some stages in the production of *nori* sheets are now mechanized. At first, a vacuum cleaner-like apparatus was used to dislodge the seaweeds from the nets, which remained in the water, and suck them into the boats. Now, the fishers use special low barges that can sail under the nets, which are lifted past a rotating clipper that resembles an old-fashioned reel lawnmower. This device carefully cuts off the seaweed blades without damaging either the nets or the holdfasts of the seaweeds. The nets slide back into the water, where the seaweeds can start to grow again if there is sufficient sunlight and nutrient material, and the freshly cut seaweeds drop into basins placed at the bottom of the boat. At the peak of the season, a new crop can be collected every ten days or so; a net with *Porphyra* is typically harvested five times. A net that is ca. 1.2 meters wide by 18 meters long can yield sufficient seaweed mass from each harvesting to make between 300 and 2,000 *nori* sheets depending on when in the season it takes place, as well as other factors that influence growth.

At the end of the day, Norio-san showed us around the *nori* factory, which belongs to the cooperative and is located on a pier in the small harbour in Kisarazu. The equipment that will be used in September and October to seed the nets with the *Porphyra* culture before they are put out in the water is stashed beside it. Outside the factory there are some open tanks filled with seawater. The newly harvested *Porphyra* is dumped into them and kept in suspension with the help of a rotor, which keeps it alive and fresh. From here, the seaweed-water mixture is pumped into the factory through a pipe to be rinsed in salt water. This removes layers of dust particles, epiphytes, and, if present, spore-bearing loose fragments, which could contribute to a discolored end product. The seaweed blades are then cut into smaller pieces, which are suspended in fresh water to make a sort of blackish-brown porridge. In turn, this is fed into a machine that measures out a precise quantity of the seaweed slurry into frames set on bamboo mats that pass by on a conveyor belt. It forms a thin, nearly square layer, with the familiar dimensions of a sheet of *nori*, that is, 19×21 centimeters.

The process is very similar to that used in paper making. Precise measuring of the quantity poured into each frame helps to ensure that the finished sheet of *nori* will be thin, of uniform thickness, and without holes. In contrast to paper making, the production of *nori* does not require the addition of any substances to bind the fibers together; the soluble polysaccharides in the seaweeds act as a natural glue. While it is still wet, the seaweed slurry on the bamboo mats is compressed with a flat sponge, which absorbs much of the water, so that the thin sheets start to take on the appearance of finished *nori*. Still on the bamboo mats, they are passed through a drying machine in which a current of air with a temperature of between 40° and 50°C completely dehydrates them. Each sheet now weighs about 3 grams. They are removed from the mats, those with holes or other defects are automatically rejected, and the finished *nori* sheets are piled up in bundles of ten. The *nori* bundles are then sealed in air-tight plastic bags in order to keep the sheets completely dry. The entire process—from the time the newly harvested *Porphyra* is placed in the outdoor tanks to the time when the finished *nori* sheets are assembled into bundles and sealed—proceeds in an assembly-line fashion. But the final step, packing the bundles into boxes ready for shipping, is done by hand.

Norio Kikuchi, the *Porphyra* researcher, pointed out that the dried *nori* sheets, also called *hoshi-nori*, in a certain sense consist of seaweeds that are still alive. Even though they are thoroughly dehydrated, the cells of the largely undifferentiated *Porphyra* fragments are biologically intact. The relatively low temperatures in the drying machine destroy neither the proteins nor the enzymes and, as a consequence, the finished product has a limited shelf life. *Hoshi-nori* of the highest quality is pitch-black and soft, has a shiny top surface, and is without holes or irregularities. The sheets produced at the beginning of the harvesting season are the best.

When *hoshi-nori* is heated, its color changes to a beautiful dark green. For example, this happens when it is toasted over a gas flame to make *yaki-nori*. Toasting, which is often accompanied by the addition of small quantities of taste enhancers such as soy sauce, changes the taste of the seaweed and the sheets become stiffer and crisper as long as they are kept bone-dry.

Back at the inn, the six *nori* fishers joined us following their NGO meeting. They are a jovial bunch and it is obvious that they share a strong working fellowship that fits in perfectly with the male-dominated clientele of the inn. Beer was served first, together with a variety of appetizers, and a large bottle of *sake* soon made an appearance. We sat on pillows on the floor, which was covered with *tatami* mats. Little by little, a large assortment of simple, delicious dishes, in which *Porphyra* and Norio-san's *nori* were, naturally, well represented, were brought to the small, low table.

Porphyra tastes best as dried, toasted *nori*, but can really also be eaten fresh just as it is. At the evening meal at Minshuku Yohei, one of the many traditional side dishes was *Porphyra* marinated in a little rice vinegar and soy sauce. It could be thought of as a Japanese counterpart to the cucumber salad served in so many countries in Europe.

At the end of the evening's get-together with the *nori* fishers of Kisarazu, I was discreetly handed a bag with ten packages of Norio-san's fine *yaki-nori*. Ten packages is equivalent to 100 sheets of *nori* and I could not help thinking that Norio-san and his *nori* fishers would have to rise at dawn the next morning and harvest at least five kilograms of fresh *Porphyra* to be able to produce 100 sheets of *nori*.

▸▸ *Porphyra tenera* cultivated on a rope in Japan in 1968.

No. 8

34.1.15

WAKAME—NEITHER TOO TOUGH NOR TOO DELICATE

Wakame is the Japanese name for the brown macroalgae *Undaria pinnatifida*, which belongs to the order Laminariales and is a type of kelp. It has an attractive, dark greenish-brown color, with blades that can grow to a width of 30–40 centimeters and a length of 60–120 centimeters.

Wakame is most commonly cultivated along the northern Japanese shorelines, especially in Hokkaido. After harvesting, the macroalga is blanched or boiled and then dried. Alternatively, it can be salted.

Wakame, which is best known from its ubiquitous presence in all varieties of *miso* soup, has a sweet, mild taste and requires a minimum of preparation. It can be cooked for a short time in soups, can be eaten toasted, or soaked in water for a few minutes and made into a salad. It breaks apart very easily if it is cooked for more than about 10–15 minutes. The sporophylls, that is, the spore-bearing blades, are regarded as a delicacy and are known in Japan as *mekabu*. These blades are sweetish and can be eaten grilled or mixed with sesame seeds and made into a snack.

▲ *Wakame* (*Undaria pinnatifida*).

In the middle of its blade, *wakame* has a tough rib, which is, however, less chewy than the one found on winged kelp described below. The blade can be cut away from *wakame* that has been rehydrated and toasted to make it crisp and quite edible.

The best known *wakame* product, which one can buy in a fish store or order at a Japanese restaurant, is seaweed salad. To make it, fresh *wakame* is placed for one minute in water at a temperature of 80°C and immediately cooled in cold water. It is then salt-preserved in a 1:3 ratio for 24 hours to dehydrate it. The blanching and salting brings out a beautiful bright green color in the *wakame*. Leftover salt is then removed, the midrib is sometimes cut away, and the seaweed is sliced into thin strips. The finished product, called *hiyashi wakame*, to which seasonings such as sesame oil and chili are added, is frozen until used.

In comparison with other brown algae, *wakame* has a lower iodine content. On the other hand, it is the species of seaweed that contains the most omega-3 fatty acids, especially EPA and its biochemical precursors. Its calcium content is almost as great as that of *hijiki*.

WINGED KELP—ALMOST AS GOOD AS *WAKAME*

Two different species are commonly called winged kelp: *Alaria esculenta*, which grows along the coastal areas of the North Atlantic, and *Alaria marginata*, which is found in the Pacific from California to Alaska. To complicate matters, the former also goes by the name of Atlantic *wakame* or wild *wakame*, even though it belongs to a completely different genus. On occasion, winged kelp is also just called kelp, using the umbrella term for large brown algae, which is reasonable given that it belongs to the order Laminariales. It has thinner, softer blades than other types of kelp and its color tends toward yellowish and olive green with a golden midrib. Winged kelp has very characteristic long spore-bearing blades (sporophylls), which branch out from the stipe above the holdfast. They resemble small wings, hence, the Latin name derived from *ala*, meaning wing, and the common English name of the seaweed. The main blade, which can attain a length of 1–3 meters, has irregular segments, especially toward the tip, which look a little like a fringe.

In many cases, dried winged kelp can be substituted for Japanese *wakame* in recipes, but it requires a longer soaking time (ca. 20 minutes). After it has been rehydrated in cold water, dried winged kelp is almost as good as fresh. It has a mild taste and can be used as a salad. The midrib is edible if it is toasted and deep frying the sporophylls brings out a taste that is reminiscent of peanuts.

Winged kelp is one of the seaweed species with the highest vitamin A content, comparable to that of spinach and parsley. It also contains a significant amount of calcium, almost equivalent to that in sesame seeds, and potassium is present in greater quantities than sodium.

BLADDER WRACK—ABUNDANT EVERYWHERE

Bladder wrack is a brown alga from the genus *Fucus*, the best known example of which is *Fucus vesiculosus*. Although it is widespread along the coastlines in virtually every part of the world and is possibly one of the more valuable of all the species of seaweeds, it is held in relatively low esteem. Some varieties of bladder wrack have traditionally been used to make a tea and it can be used in the same way as *konbu*. It is not very often utilized as food for humans, which is a shame, as the youngest and outermost shoots of the seaweed are extremely tasty.

Fucus is commonly dried and sold in two forms: as granules or as small branches, which are a beautiful deep green or brownish color and have their dehydrated air bladders attached. It has a strong taste of iodine and is generally very salty, both indicators that bladder wrack has a chemical composition that mirrors that of *konbu*. It is good in cooked dishes, soups, or sprinkled on salads.

Edible marine algae

▲ Winged kelp (*Alaria esculenta*).

▲ Bladder wrack
(*Fucus vesiculosus*).

HERB. E. GEORGE.
Presented by the London County Council, 1915.

Whitby
June 1866.

▲ Dried specimen of sporophylls on winged kelp (*Alaria esculenta*) from the collections of the Natural History Museum in London.

alaria esculenta

Sea palm—a rare delicacy

As the name implies, sea palm (*Postelsia palmaeformis*) is a brown alga that resembles a small palm tree. Its habitat stretches along the west coast of North America from central California to British Columbia. It grows in the upper intertidal zone, where it is often buffeted by heavy surfs. When one first comes across sea palms at low tide, one is surprised to see a small erect palm, anchored firmly to a rock by a broad, sturdy holdfast, with a thick, short stipe and masses of blades hanging down from the top. The blades are narrow and can be up to 40 centimeters long. They are olive green and have a characteristic grooved pattern running along their length. As they age, sea palms take on a more yellowish-brown shade.

Sea palms are regarded as a threatened species and individuals are forbidden from gathering them in many jurisdictions. In California, their harvest is permitted for limited commercial use.

The blades of sea palms are among those with the greatest fiber content, about 65%, which is considerably more than dulse. For this reason, many people think that they promote digestive health. Only the blades are eaten; they can be dried and toasted and consumed as snacks. Other methods of preparation involve cooking and marinating. When rehydrated, sea palm blades swell up like cooked noodles and, like other delicate brown algae, can be used in a salad.

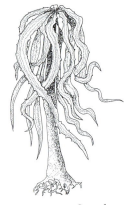

▲ Sea palm
(*Postelsia palmaeformis*).

Arame—totally mild and sweet

Arame (*Eisenia bicyclis* or *Eisenia arborea*) is a brown alga harvested along the Pacific coast of Japan, where it thrives best in relatively cold water. It is first sun dried, then steamed or boiled for several hours, cut into strips, and dried once again.

Dried *arame* has a very dark color, but it turns brownish when it is soaked in water. This also causes it to double in volume. It takes only about five minutes to soften it and it loses its flavor if it is left in the water too long.

In terms of taste, *arame* resembles *hijiki*, but it is milder and less salty. Probably it is the sweetest of all the species of seaweed that are commonly eaten. It is especially well suited for salads and soups, but can also be marinated. After it has been cooked for only a few minutes, it becomes very soft and swells up. It contains a fair amount of calcium, which is present in chelated form, that is, bound to an organic or amino acid, which permits the body to absorb more of the mineral.

▲ *Arame* (*Eisenia bicyclis*).

Hijiki—a bit special and a little poisonous

Hijiki (*Sargassum fusiforme*, formerly *Hizikia fusiforme*) is a brown alga that is greenish-brown and has a bushy shape with narrow blades. It grows in the wild in shallow waters along the coasts of Korea, China, and Japan and in cultivated form in Korea. It is harvested early in the year and the blades are sun dried, then boiled in copious amounts of water, and dried again. In dried form it is completely black.

One of the reasons for cooking *hijiki* is that it contains considerable quantities of arsenic, a poison that attacks the nervous system. Boiling releases a proportion of the arsenic into the water, but some of it remains. Another reason for boiling *hijiki* is that it contains a number of bitter color substances, polyphenols, which are also released into the cooking water. As a consequence of the boiling, *hijiki* looses some of its appealing black color. In order to preserve the dark color in the cooked seaweed, it is often boiled with another brown alga, for example, *arame*, which imparts some of its dark color to *hijiki*.

Due to its arsenic content, government food safety agencies in a number of countries have issued warnings against eating *hijiki* in large quantities. The Japanese have been eating *hijiki* for centuries, however, seemingly without any serious problems due to arsenic poisoning. It is possible that the metabolism of the Japanese, who eat a great deal of *hijiki*, over time has adjusted to excrete arsenic and, consequently, they suffer no ill effects. Another explanation may be that arsenic, which forms organic compounds such as arsenoriboside, is less poisonous in this form than inorganic arsenic, the form on which the government agencies have based their evaluation of a given product's suitability as food.

Dried *hijiki* is soaked in water for 10–20 minutes, in the process swelling to three to five times its original size. This water is discarded and the alga can then be boiled or simmered for at least 30–40 minutes to make it sufficiently tender.

Hijiki has a pleasant firm texture and a mild, nutty taste, which may also come across as a bit insipid. It can, therefore, benefit from the addition of a dash of soy sauce. It pairs well with ingredients that are somewhat sweet, especially if it has been cooked with a little rice wine, *mirin*, and is particularly good in simmered dishes. *Hijiki* also combines well with green asparagus and can be sprinkled on top of a green salad, where its black color forms a beautiful contrast. For the same reason, it is decorative when mixed into a salad of raw carrots.

Edible marine algae

▲ *Hijiki* (*Sargassum fusiforme*).

89

Hijiki is one of the species of seaweeds that has the highest content of omega-3 fatty acids (EPA), about two to three times as much as *nori*. It also contains a reasonable quantity of iodine and is richer in calcium than any of the other seaweeds. A tablespoonful of *hijiki* has as much calcium as a glass of milk. Over and above all of its benefits, *hijiki* is among the best tasting of the marine algae.

DULSE—THE ROBUST RED ALGA

A rich cultural history is associated with the red alga dulse (*Palmaria palmata*), which has been eaten for centuries by humans living along coastlines around the world, especially those of the northern Atlantic and Pacific Oceans. This seaweed has dark red or purple blades that are of varying widths and can be up to half a meter long. It has been given the name *Palmaria* because the blades are sometimes split into sections that look like the fingers on a hand. In most places, dulse grows in the lower part of the intertidal zone and can, therefore, easily be gathered when the tide has gone out. It is usually harvested in the spring months. In some cases, the fresh dulse is preserved in salt, but it is usually dried and sometimes chopped up into pieces or ground into coarse granules. In the course of the drying process, salts may seep out onto the surface and show up as white spots on the reddish-purple blades. Of all the marine algae, dulse probably has the taste that is most agreeable to the Western palate.

When fresh, dulse can be eaten as a salad, possibly after being soaked in fresh water, which causes its cells to burst. It is often too tough to eat raw, but drying makes it easier to chew and brings out its pleasant salty and nut-like taste. Dulse should not be cooked as this causes it to break up.

Dulse has a relatively high protein content, typically more than 20%, which is greater than that of foods such as chicken or almonds, although it lacks the essential amino acid tryptophan. It also contains a significant amount of potassium, but in comparison with other seaweed species has only a little fat and less iodine than the large brown algae. The carbohydrates found in dulse are in simpler form than in other species of seaweeds, which results in a more interesting taste. Dulse is also rich in iron, having almost as much as bladder wrack.

In modern cuisine, dulse is incorporated into bread, fish and vegetable soups, and fish dishes, or simply eaten toasted as a snack that goes well with a dark beer or an aperitif. It can be fried to a crisp in a little oil or butter and used as an agreeable substitute for fried bacon. Because it has a somewhat

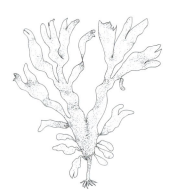

▲ Dulse (*Palmaria palmata*).

St. Columba's monks were eating dulse on Iona in the Hebrides in the 6th Century CE: "Let me do my daily work, gathering dulse, catching fish, giving food to the poor." From the Celtic Psaltery, attributed to St. Columba (521–597).

A particularly tasty variety of dulse smoked over applewood is produced in Maine. In Nova Scotia, Canada, one could earlier buy a cultivated variety of dulse sold as 'Sea Parsley'.

▶ ▶ Dried dulse. The white spots are made up of flavorful salts that have seeped to the surface during the drying process.

sweet taste, dulse is a good complement for root vegetables and corn. It can also be crushed and sprinkled on salads and vegetables. When toasted, dulse loses some of its red color and looks brownish.

Edible green algae

Sea lettuce (*Ulva lactuca*) is a green alga that consists of large, broad blades that can be up to a meter wide. The ruffled blades, which are attached to the holdfast without a stipe, are pale green and very thin, only two cells in thickness. Sea lettuce is very common and found along most coastlines, although it shows a preference for nutrient-rich environments. It thrives well in seawater that has low salinity, but can become a nuisance in those areas where large quantities of nutrients wash into the sea. Its growth is facilitated by its ability to absorb organic nutrients, such as glucose, directly from the water.

▲ Sea lettuce (*Ulva lactuca*).

► *Ao-nori,* dried green algae (*Ulva lactuca* and *Ulva prolifera*).

Some seaweeds that were formerly identified as *Enteromorpha* have now been reclassified as species of *Ulva*. Some of these, such as green string lettuce (*Ulva linza*), have more elongated blades, which in some species, for example, gut weed (*Ulva intestinalis*), fold around themselves like a narrow tube.

Compared with other types of seaweeds, sea lettuce is rich in protein and low in fat. It contains only a little iodine, but is abundant in iron.

In Japan, another genus of green alga, *Monostroma*, which resembles sea lettuce but is composed of blades that are only a single cell layer in thickness, is also cultivated and harvested. Both *Monostroma* and *Ulva* are sometimes treated like *Porphyra* and made into a cheap imitation of *nori*, known as 'poor man's *nori*'.

As is implied by its name, sea lettuce can be eaten raw as a salad. It has a delicate, soft texture, but a strong taste. In its dried form, sea lettuce is tough, but becomes soft after being soaked for a few minutes in water. Various species of *Ulva* and *Monostroma* can be dried and toasted or used as an additive to

▲ Green string lettuce (*Ulva* sp., formerly known as *Enteromorpha* sp.).

enhance the taste of warm dishes, soups, and salads. The dried algae are used to prepare small flakes, known as *ao-nori* (green *nori*), which are sprinkled over warm rice as a seasoning.

AGAR—COMPLETELY WITHOUT TASTE

Agar, the Malay name for red algae, also called agar-agar and *kanten*, is an extract of red algae, sold in granular or powder form or as flakes or long strips. It is widely used as a gelling agent and rarely eaten on its own. In its pure form, agar is a tasteless and odorless polysaccharide, which normally contains the same minerals, proteins, and vitamins as the red algae from which it is derived, but in lesser quantities.

Agar is light in color, semi-translucent, and very brittle when dry. When soaked it absorbs water and, in contrast to gelatine, which is prepared from animal and fish skin and bones, it must be allowed to swell up completely before the water is warmed to a temperature above its melting point of 85°C. It can then be used as a gelling agent as it cools.

It is said that *kanten* was discovered by accident. For more than a thousand years, the Japanese have eaten a dish called *tokoroten*, which is prepared by first boiling a type of red alga, *tengusa* (*Gelidium amansii*), and then letting the mixture stiffen. At some point toward the end of the 17th Century, leftover *tokoroten* was thrown away outside on a freezing cold day. When it was found later, it had become a dry, whitish solid. Water was added to it, it was boiled again and then dried, yielding an even whiter, finer mass with a better taste. The result was given the name *kanten*.

SPIRULINA—EDIBLE BLUE-GREEN MICROALGAE THAT PACK A PUNCH

Spirulina is the common commercial designation for two species of blue-green microalgae, *Arthrospira platensis* and *Arthrospira maxima,* which technically speaking are not real algae but cyanobacteria. Spirulina floats freely in water as multicellular filaments, which form a left-handed spiral coil. It grows naturally in tropical and subtropical fresh water lakes with a low pH reading and is an important aquaculture crop in North America, Australia, New Zealand, and Asia. It produces a large biomass that can be filtered out of the water, dried or freeze-dried, and made into a powder or compressed into pills.

In order for microalgae to be used for human nutrition, it is necessary to break down their cell walls, as they consist of indigestible carbohydrates.

Edible marine algae

▲ Agar, also called *kanten*.

Agar in Japanese *haiku*

matsu yori
mo furuki kao
shite tokoroten

his face
older than the pine
sweet jelly

Kobayashi Issa (1763–1827)
(translated by David Lanoue)

93

► Microscopic image of Spirulina, the blue-green microalgae *Arthrospira platensis* and *Arthrospira maxima*. The microalgae assemble themselves into long filamentous spirals.

It is thought that the reason that microalgae contain so much DHA is that this enables them to adapt quickly to new environments. From DHA the organism can synthesize the signal molecules (eicosanoids) that are important for growth. Blue-green microalgae are the most important sources in the food chain of the DHA and EPA that end up in fish and shellfish and, ultimately, in humans.

Growing Spirulina has been proposed as the ideal way to produce food efficiently during space voyages of long duration.

Hence, one could normally derive only a little benefit from eating raw blue-green microalgae. One can improve the accessibility of the useful nutrients by drying the microalgae in the sun or a rotating drum, or freeze-drying, all of which help to destroy the cell walls. Fortunately, the cells walls of Spirulina are easier to break down than those of other microalgae, such as *Chlorella*.

Spirulina is one of the most carefully studied functional foods. Dried Spirulina typically contains over 60% protein, 7–8% fats, and about 15% carbohydrates. Almost 60% of the fats are unsaturated and about half of them are polyunsaturated. The content of omega-3 fatty acids (DHA and EPA) is almost double that of omega-6 fatty acids (especially gamma-linolenic acid and AA). It is noteworthy that these microalgae are particularly rich in DHA, actually they have about twice as much DHA as EPA. In this respect they are distinctly different from the macroalgae, which are characterized by a preponderance of EPA and virtually no DHA.

Because Spirulina contains approximately as much potassium as it does sodium, it is a good source of salt for those who are at risk of high blood pressure. It also has an abundance of chlorophyll and carotenoids, which are important antioxidants.

In relation to its mass, Spirulina is a tremendous source of food energy and nutritional elements. There are about 1500kJ (360 calories) per 100 grams of dry weight and no other food has as great a concentration of proteins. So it is hardly surprising that Spirulina is called a superfood and is so prominent on the shelves of health food stores. Nutritional experts recommend an intake of ca. 5mg of Spirulina twice daily.

The taste of seaweeds

TASTE AND TEXTURE

Humans have probably never eaten the ordinary varieties of seaweeds simply because they were palatable, but more likely because they were readily available in coastal areas. On the other hand, there is little doubt that it was on account of their taste that the more delectable species came to be so highly prized and early on found their way to the tables of the nobility and the imperial courts in China and Japan.

In today's cuisine, seaweeds are valued not only for their taste but also for their texture. Taste is very dependent on how the algae have been treated and whether they are being eaten raw, dried, toasted, or cooked. In addition, the aromas that are released by the seaweeds play a major role in the taste sensations they evoke. Seaweeds contain small peptides, which are the reason why some seaweeds, especially *konbu* that has been cooked for too long, may taste bitter.

When toasted, a number of seaweeds, particularly *nori*, *wakame*, and dulse, become so crisp that they break up into flakes in the mouth before they have a chance to absorb any moisture. They can easily be crumbled in the hand and sprinkled over a salad or a bowl of rice. This crunchy quality is an important element of the taste sensation. Coarser varieties of seaweeds, for example, *konbu*, which are cooked and presented in dishes such as soups or salads, have the mouthfeel of something that is both soft and elastic.

The Japanese particularly value seaweeds for their texture and use three different terms for this property: *kuchi atari* (palatability, mouthfeel), *shitazawari* (tongue feel), and *hagotae* (crunchiness, tooth resistance).

Seaweed products that are used as gelling agents, namely, agar, carrageenan, and alginate, have no taste and are, therefore, suitable for use in desserts where it is desirable for these additives to be taste-neutral.

As we will see later, brown and red algae, in particular, are used to add flavor to different types of food, especially soups, salads, and sushi. Toasted seaweeds, in which a salty taste combines with overtones of nuts, smoke, and iodine, are eaten as snack foods.

It has often been said that seaweeds offer our palates a veritable smorgasbord of taste impressions, on any occasion, in any dish: *konbu* contributes an intense and robust *umami* and iodine flavor; *hijiki* adds texture and a wealth of sweet taste sensations; dulse is redolent of the ocean itself; *wakame* is chosen for its mild, delicate taste; *nori* is like a fresh sea breeze.

▲ Sea grapes (*Caulerpa lentillifera*) are green seaweeds that resemble bunches of small grapes. They are eaten fresh as a salad, e.g., in the Philippines, where they are known as 'green caviar', and in Japan, where they are called *umibudou* (sea grapes). Sea grapes have a crispy texture and a refreshing juicy and succulent mouthfeel. The particular sensation of the grapes bursting in the mouth is valued by the Japanese, who even have a special expression, *puchi-puchi*, to characterize it.

SEAWEEDS AND *UMAMI*

The taste of seaweeds is closely tied to what is known as the fifth taste or *umami*. In Asia, one has for many years talked about five different types of taste. In addition to the four well-known ones—sour, sweet, salt, and bitter—there is a fifth taste, *umami*, which in Japanese means something along the lines of savory or delectable.

That *umami* is a distinct taste in the sensory physiological sense was established scientifically in 2000, when the first specific taste receptor that is able to recognize the key substance imparting the *umami* taste was identified. This substance is monosodium glutamate, MSG, the sodium salt of glutamic acid, an amino acid found in great abundance in seaweeds. MSG is sometimes referred to as the third spice, the first two being salt and pepper. It is used widely as a taste enhancer in Chinese cuisine. MSG is also found in cured ham, Parmesan cheese, mature tomatoes, as well as fish and soy sauce. Hence, terms often used in English to characterize *umami* taste are brothy and meaty.

Brown algae such as *konbu* have a particularly high MSG content. This substance is released when the seaweeds are softened and heated gently in water. Normally one should not wash dried seaweeds before using them, as tasty minerals and amino acids have often seeped out onto their surfaces. Their taste can, however, also depend on how they are treated. When making soups, for example, the seaweeds should not be boiled for very long or, better yet, not at all, as prolonged cooking can bring out a taste that is too strong and fish-like.

Other substances, which are derived from nucleic acids that are dissolved within the cells of some seaweeds, in particular *nori*, are also sources of the *umami* taste. These nucleic acids, especially inosine monophosphate and guanosine monophosphate, are formed when the cells break down ATP (adenosine triphosphate), the energy storing molecule, in order to obtain the energy they require to do their work. Nucleic acids impart a sweetish taste to seaweeds and these taste substances can be transferred to the animals that eat the algae. Sea urchin roe is sought out for its sweetness and *umami* taste, which is due to the inosine monophosphate that sea urchins ingest when they graze in kelp forests.

The reason why seaweeds contain many of these substances that draw out the *umami* taste is that they help to maintain the correct osmotic balance in the seaweed cells so that they do not burst when they are exposed to the surrounding, often very salty, seawater. Consequently, seaweed species from more saline oceans have a stronger *umami* taste. Another substance that

Seaweeds & human nutrition

In 1908, a Japanese chemist, Kikunae Ikeda (1864–1936), identified MSG as the chemical compound that is responsible for the savory taste of the traditional Japanese soup stock, *dashi*. This broth is made from a warm aqueous extract of *konbu* and a conserved fish product called *katsuobushi*. *Konbu* turned out to be very rich in MSG, which makes up about 2–3% of its dry weight. Ikeda coined the term *umami* to describe the taste of MSG, from the Japanese *umai* (delicious taste) and *mi* (essence).

Of the many different variants of Japanese *konbu*, *ma-konbu*, *rausu-konbu*, and *rishiri-konbu* are considered to be the best bases for *dashi* and they yield a very light *dashi* with a mild and somewhat complex taste. *Ma-konbu* is the *konbu* with the largest amount of free MSG, 3200 mg/100 g, whereas *rausu-konbu* has 2200 mg/100 g, and *rishiri-konbu* has 2000 mg/100 g. The lower quality *hidaka-konbu* has 1300 mg/100 g.

▸▸ Dried *arame* and bladder wrack.

96

probably also helps to regulate osmosis is the sugar alcohol, mannitol, working together with a number of polysaccharides. The brown alga sugar kelp (*Saccharina latissima*) is so named because, when dried, it exudes mannitol, which is sweet. Mannitol is often deposited on the surface of dehydrated sugar kelp as a white powder.

SMOKE AND SULPHUR

Some seaweed species, especially *konbu* and a number of other brown algae, taste strongly of iodine. Toasted bullwhip kelp and giant kelp impart a distinctive combination of taste impressions that are a mixture of smoke and iodine.

Seaweeds can also have flavors and emit odors that are due to sulphur compounds, especially dimethyl sulphide and methyl mercaptan, which are formed by microorganisms living on their surface. Here they break down the chemical substance DMSP (dimethylsulfoniopropionate) that also helps the cells to regulate osmotic pressure. In small quantities, dimethyl sulphide gives off a pleasant smell of the sea, but in large amounts it is very unpleasant and is an indication that the seaweed is in the process of decomposing.

Seaweeds as salt

IF YOU HAVE NO SALT, YOU HAVE NOTHING

In earlier times, salt was a precious commodity, which was sometimes referred to as white gold and was used by the Vikings as a medium of exchange. It was valuable in an age when salting meat and fish was one of the only ways to preserve them. According to the Roman historian Pliny the Elder, the legionnaires were paid in salt, a concept that survives in the English word for payment for work, salary, and its cognates in other languages, all derived from the Latin *salarium*. Neither bread nor meat is enjoyable without salt. There are many proverbs about how important salt is for human life, among them the succinct Nordic saying, "If you have no salt, you have nothing."

Most ordinary white table salt is now derived from underground salt domes, formed millions of years ago. Called rock salt, it is almost pure sodium chloride (NaCl), about 99.3%, and has a smattering of other salts and minerals. Archaeological research has recently confirmed that evidence of extensive exploitation of the Duzdagi salt deposits in Azerbaijan indicates that it is the oldest known salt mine in the world, dating from about 4500 BCE. But before the advent of large machines and the development of brine extraction methods, salt mining was expensive and dangerous.

Consequently, the major source of salt throughout the Middle Ages was the sea. Coastal peoples were always able to obtain what is known as grey salt. In warm climates this was done by allowing seawater to evaporate in the sun and in colder areas by boiling the water over an open fire. Sea salt consists of 90–96% sodium chloride and ca. 4–7% water, as well as varying amounts of other salts and minerals, depending on how the salt was produced.

Seaweeds as salt

SEAWEED ASH SALT AND SMOKED SALT

There is, however, another source of salt in the oceans, namely, seaweeds, which have a greater concentration of salt than the water that surrounds them. Salt can be produced quite simply by burning dried seaweeds to ash. This results in a black residue, which contains sodium chloride (NaCl), sodium carbonate (Na_2CO_3), and potash (K_2CO_3). The ash is placed in large tubs, mixed with seawater, and the mixture is heated.

As the water temperature is raised gently to over 80°C, the water gradually evaporates and the salt concentration becomes so great that crystals start to form on the surface. Sand particles and other impurities fall out to the bottom of the vessel and the dark salt, which contains a large amount of ash, is skimmed off the top.

It is likely that this technique goes back to ancient times and we know that it was practiced along the Danish coasts, especially in northern Jutland and nearby islands. This local source was probably vitally important when the trade in the finer white rock and sea salts from the south and east was disrupted by events such as the Napoleonic wars. Even if this seaweed ash salt was of less value and lower quality than imported salts, it could always be used for food preservation.

The traditional method of making it was to dry the seaweeds, burn them in a pit in the sand on the beach, and gather up the ash. The rest of the process was energy-intensive, using dried seaweeds and wood, and this was not without some harmful environmental side effects. For example, the small Danish island of Læsø, which was well-known in the Middle Ages for its salt industry, was so thoroughly denuded of trees by the mid-1600's that the king forbade the use of wood for salt production. In other areas of the coastline, the extensive deforestation allowed storms to drive the sand inland, ruining a great deal of valuable agricultural land.

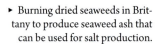

▶ Burning dried seaweeds in Brittany to produce seaweed ash that can be used for salt production.

Seaweed ash salt sometimes has a smoky taste, which is caused when residue from a fire finds its way into the evaporating tub. Even more interesting flavors can be created if the ash-seawater mixture is warmed up and undergoes evaporation by plunging red-hot stones into a wooden tub. Bits of seaweeds that have been carbonized by contact with the stones and left in the mixture can provide an intense smoky flavor. Smoked salt can also be produced in the time-honored way by exposing the salt over a long period of time to smoke from different types of wood, such as juniper and oak, which impart their own subtle tastes.

Because of their interesting taste nuances, seaweed-based salts are well suited for seasoning and enhancing the flavor of soups, fish, meat, and bread. Apart from this, however, they have beneficial side effects that can improve overall health. The greater proportion of potassium in relation to sodium found in seaweed ash salt can help to lower blood pressure, a boon for those who suffer from hypertension.

In addition, salt made from brown algae contains a great deal of iodine, which is vital for the proper functioning of the thyroid. An adult needs a daily intake of about 150 µg of iodine in order for this gland to be able to synthesize vital hormones, including those that control metabolism.

As this essential element is often lacking in the diet, many countries have made it mandatory to add iodine to common table salt. This is not required for salts containing seaweed residues, however, as they naturally already contain large quantities of iodine.

Seaweeds, wellness, and nutrition

FOLK TRADITIONS AND THE HEALTH MOVEMENT

For many millennia, the original inhabitants of all continents, especially those in coastal areas, have availed themselves of seaweeds for food and medicine. In the last few decades, numerous health movements have embraced seaweeds and marine algae and touted their beneficial aspects. Some of these claims are well founded, at least from a scientific point of view. In other cases, probably the majority, the supposed positive effects are supported by references to tradition and experience. There is a real need for research in this field.

Nevertheless, there are some interesting observations that are hard to avoid. It is well-known that people who inhabit places where consumption of substantial quantities of seaweeds and other marine foodstuffs is the norm have a lower incidence of cardiovascular disease and high blood pressure. They also tend to live longer. This latter fact was already pointed out in 1927 by the Japanese professor Shoji Kondo of Tohoku University, who was investigating the correlation between lifespan and diet in various areas of Japan. His findings, later corroborated by more recent studies, were that on those islands of southern Japan where the consumption of seaweeds is high, life expectancy, especially that of women, is generally longer. These population groups also had a low calorie intake, ate less rice, and used less salt in their food than those in other parts of Japan.

Throughout the ages, and increasingly in connection with a variety of modern health movements, beneficial properties and the ability to cure all sorts of ailments, from digestive problems to cancer, have been ascribed to seaweeds, algae, and products made from them.

WHAT HAVE WE ALREADY LEARNED?

In general terms, it can be said that a varied diet that includes a proportion of seaweed products, for example, up to 10% as in Japan, promotes wellness. This is due principally to the high concentration in marine algae of important minerals and vitamins. These minerals in seaweeds are in what are known as chelated and colloidal forms, which enhance their bioavailability in the body. Seaweeds are also a good source of proteins and essential amino acids.

In addition, marine algae have a much greater fiber content than vegetables and fruits, as they are largely composed of both soluble and insoluble dietary fiber. Because dietary fiber is indigestible, it contributes no calories, but instead enhances digestive function by absorbing water and thereby easing the passage of food through the intestines. Soluble fiber, in particular, slows the rate at which carbohydrates are absorbed, which helps to lower blood sugar levels, a potential advantage for diabetics.

Also, the alginates found in the cell walls of brown algae appear to have the beneficial property of being able to bind heavy metals and radioactive isotopes such as strontium-90 and iodine-131. It has been shown that sodium alginate reduces the uptake of strontium in the intestines by a factor of seven. One drawback, however, is that the fiber can also retain important micronutrients, preventing them from being absorbed in the digestive tract.

Seaweeds, especially the brown algae, contain chlorophyll, carotenoids, vitamins, sterols, and phenols, which have well-documented antioxidant properties that can help to ward off cardiovascular disease and some types of cancer. Studies currently under way indicate that seaweeds also protect against viral diseases and bolster the immune system.

The relatively high level of unsaturated and essential fatty acids, especially the omega-3 fatty acid EPA, is a good reason for eating seaweeds. It is widely accepted that omega-3 fatty acids counteract cardiovascular disease and reduce the risks of severe aftereffects resulting from blood clots and, in addition, inhibit the build-up of cholesterol in the arteries.

Omega-3's also play an important role in the development of the brain and the nervous system, particularly *in utero* and in the first year of life and, presumably, in keeping them in good working order well into old age. Some studies indicate that a good intake of omega-3 fatty acids, with an appropriate balance between omega-6 and omega-3 fatty acids, can reduce the risk of psychological illnesses such as bipolar disorder, schizophrenia, and certain neurodegenerative diseases.

Marine algae strike a nutritionally perfect balance between omega-3 and omega-6 fatty acids.

The above needs to be looked at in the light of two circumstances. First of all, the diet in Western countries is out of balance with respect to its proportion of omega-6 to omega-3 fatty acids, as the consumption of omega-6 fatty acids is far too high. Secondly, a drastic growth in the incidence of mental illness is coming our way in the slipstream of the widespread lifestyle diseases that have already plagued the 20th Century, that is to say, cardiovascular disease, hypertension, obesity, type 2 diabetes, and cancer. Much research indicates that this can be attributed to dietary factors, which include too little fiber, too few omega-3 fatty acids, and too many calories, including those from refined carbohydrates.

Utilizing seaweeds to impart a salty taste to foods has the nutritional advantage that its disproportionately high potassium content does not interact with the sodium in the bloodstream and, therefore, does not elevate blood pressure as does ordinary table salt.

One of the most important contributions of seaweeds to wellness might turn out to be the combination of effects brought about by dietary fiber and seaweed sterols, in particular fucosterol. Apart from the beneficial way in which soluble and insoluble fiber work in the digestive system and increase the feeling of satiety, they seemingly also positively affect the circulatory system. Fucosterol brings about a decrease in the cholesterol level in the blood.

The mechanisms responsible for all this are not known in detail, but research has shown that it is possible that several different ones are involved. In the first place, fucosterol depresses the biosynthesis of cholesterol. Secondly, it seems that fucosterol elevates the amount of so-called good cholesterol (HDL or high-density lipoprotein) and reduces the amount of bad cholesterol (LDL or low-density lipoprotein). Thirdly, dietary fiber content comes into play. For example, the cholesterol level in the liver is reduced when alginates moving through the intestines bind with bile salts and, with them, a quantity of cholesterol, all of which are then excreted from the body.

Furthermore, it turns out that several of the polysaccharides in seaweeds also lower the level of LDL, free cholesterol, and triglycerides in the bloodstream. The combined effects of dietary fiber and seaweed sterols on the entire complex of problems associated with cholesterol and its transport within the body are, therefore, positive and can help to lower the risk of cardiovascular disease.

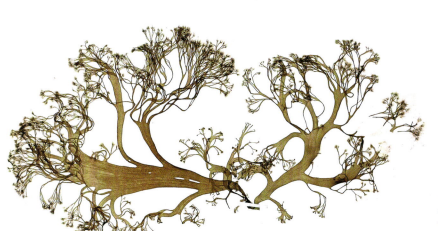

◄ Dried specimen of the brown alga *Cutleria multifida* from the collections of the Natural History Museum in London.

It has also been shown that fucosterol has another important function in that it inhibits the formation of blood clots by acting as an anticoagulant. The mechanism is that fucosterol enhances the action of a factor (plasminogen-activating factor) that, through a series of biochemical steps, results in increased formation of plasmin, an enzyme that helps to dissolve blood clots. It is possible that it acts in synergy with the effects produced by the seaweed polysaccharides, especially fucoidan, most commonly found in the brown algae *konbu* and *wakame*. Fucoidan stimulates the formation of heparin, a naturally occurring anticoagulant, which allows the body's own mechanisms to break down clots.

Throughout the 1980's and 1990's, the hunt was on to find specific biologically active substances in seaweeds. It has been estimated that in the decade between 1977 and 1987, 35% of all newly discovered natural substances came from seaweeds and algae. While these efforts continue, they are now directed more intensely toward marine microorganisms. Nevertheless, further research is likely to identify bioactive substances in seaweeds that have special health-promoting or medicinal properties. It is, however, not obvious that these properties should be attributed to only one or a few substances, but rather to the particular combination in which they occur in seaweeds, which quite possibly allows them to form a synergistic relationship.

In summary, it would appear that, generally speaking, seaweeds as an entity with all of their constituent substances have a predominantly positive impact on human health.

Without doubt, seaweeds are among the most obvious candidates for the role of nutraceuticals, that is, foodstuffs with potential for use as pharmaceuticals.

SEAWEEDS AND CALORIES

Marine algae are often promoted as low-calorie, protein-packed diet foods. This is correct in the sense that they contain only a little fat and few convertible carbohydrates and are so rich in fiber that they are filling. Comparatively smaller portions are called for than of some other foods that are fattier and have more convertible carbohydrates.

A team of French scientists has recently found that a microorganism that can enzymatically break down a common polysaccharide (porphyran) found in *Porphyra* is present in the gut microbiota of Japanese, but not in those of North Americans. Since *nori* made from *Porphyra* is a key ingredient in sushi, this effect has been dubbed the 'sushi factor'. Presumably, the gut microorganism must have acquired the ability to produce these enzymes at some point in the past by gene transfer from marine bacteria ingested in the food—a case of genetic pot luck.

The caloric content of dried seaweeds is comparable to that found in other common foods. It varies by species, but is typically in the range of 500–1000kJ (120–240 calories) per 100 grams of dried weight, which is equivalent to the calorie count for roast chicken. This means that raw seaweeds or dried seaweeds that have been rehydrated completely in water have approximately ten times fewer calories by weight than roast chicken.

IODINE AND METABOLISM

Iodine is the one component in seaweeds that can be said to be the most important for our health. Seaweeds, especially brown algae, have probably from the earliest periods in human evolution been either the direct or the indirect sources of iodine, on which the functioning of our brain and thyroid gland depends. The thyroid gland synthesizes two indispensable iodated hormones, thyroxine and triiodothyronine, which play a vital role in regulating the body's metabolism. Consumption of a large quantity of iodine can speed up the metabolic rate and possibly lead to weight loss. It is thought that iodine's only function in our body is in connection with the thyroid gland.

There is an interesting similarity between seaweeds and the thyroid gland in that they both accumulate iodine. The thyroid gland takes up iodine from the bloodstream and seaweeds from seawater. In the thyroid gland there is a special molecular pump that is responsible for the transport of iodine across the cell membrane, which is carried out in conjunction with that of sodium. In seaweeds the transport mechanism is not known, but it is much more efficient, especially in brown algae, as it can concentrate iodine in the cells up to a level that is 100,000 times higher than in the surrounding seawater. Likewise, seaweeds are better at retaining iodine than the thyroid gland.

It is thought that about 200 million people around the world are suffering from iodine deficiency, which is due to local factors such as a low incidence of iodine in the soil, the drinking water, and the foods they eat. The populations of mountainous regions of India, South America, Southeast Asia, and Africa are affected. Generally, most foodstuffs, apart from marine fish and shellfish and, of course, seaweeds and other marine algae, have a low iodine content. Eggs, milk, and, to a lesser extent, meat and nuts are moderately good sources of iodine. It is, nevertheless, quite difficult to ingest sufficient iodine if one does not eat a diet that includes a certain number of ingredients from the sea. To compensate for this, regulatory agencies in many countries have, since the 1920's, mandated that 0.02% iodine be added to common table salt, an amount sufficient to resolve the problem.

Iodine content varies surprisingly greatly in the different types of marine algae. Laboratory measurements have determined that it can range from 16 µg to 8 mg per gram of dried seaweed. The quantity also depends on where the seaweeds were harvested, how they have been stored, and how they are treated as foodstuffs.

Normally, seaweeds are stored in dried form and as long as they are kept in an airtight package the iodine content is preserved. As soon as they are exposed to air, especially damp air, the seaweeds can quickly lose up to half of their original iodine content. When seaweeds are boiled, the greater part of the iodine, in some cases up to 99%, is lost into the cooking water after 10–15 minutes. This water should then be discarded and not used for cooking. Iodine loss is reduced to about 20% by steaming the seaweeds and, even better, to 6% by toasting them.

Iodine deficiency leads to a weakening of thyroid function and can lead to metabolic diseases such as goitre, chronic fatigue, and, in the worst cases, to stunted growth and mental retardation, especially among children. This condition is also frequently linked to obesity.

The thyroid gland of a normal adult contains about 8 mg of iodine, of which ca. 1–2 µg per kilogram of body weight is utilized each day. For a person weighing 75 kilograms, this translates into a need for a daily intake of 150 µg of iodine. The requirement can be greater if the iodine is displaced by other substances, especially chlorine, which is added to drinking water in some places. Other problems can be caused by the radioactive isotope iodine-131, which may be released accidentally by nuclear power plants or used in radiotherapy. The iodine-131 accumulates in the thyroid and later in the reproductive organs. Iodine tablets or iodine derived from kelp can help to reduce the uptake of radioactive iodine by the thyroid gland. They constitute an essential part of the protection against radiation poisoning that could result from an accident at a nuclear plant or in conjunction with the explosion of a nuclear weapon. Evidence of this end use could be traced through the increase in the sales of kelp products following the disaster at the nuclear power plants at Chernobyl in 1986 and Fukushima in 2011.

As in most other foods, iodine in seaweeds occurs in the form of iodide (I^-), which is extremely water soluble. The total amount depends on the species. A certain proportion, about 10–20%, is in the form of iodate (IO_3^-) and a little, possibly as much as 10%, is found as organically bound iodide.

▸▸ The red alga *Clado-donta lyallii*, collected on the Falkland Islands in 1910.

Spanish Point Island Red Cove

SEAWEEDS AND CANCER

Although seaweeds and seaweed products are sometimes promoted by wellness experts and therapists as anti-cancer agents, there is, unfortunately, no strong scientific basis for such claims. Nevertheless, there are a large number of interesting studies that indicate that incorporating marine algae into the diet can suppress cancer in cell cultures and in laboratory animals such as mice. It has also been proven that *hijiki*, to take one example, stimulates the proliferation of human lymphocytes (white blood cells or T-cells), which strengthen the immune system and the body's own ability to attack tumors.

One of the bioactive substances in marine algae, which may act as an anti-cancer agent, is fucoidan, a polysaccharide found in the brown algae *konbu* and *wakame*. Its effectiveness seems to be linked to its sulphate content. Fucoidan works on the cancer cells by inducing them to commit suicide in a programmed, controlled manner (technically known as apoptosis). Some of the seaweed fats (glycolipids) have also been shown to be able to suppress tumor growth. Neither of these effects is, however, sufficiently convincing or significant to form a basis for the treatment of cancer with seaweed products. Nevertheless, it is possible to posit that a diet with a certain seaweed content can possibly have a positive, preventative effect.

An unforeseen finding is that another of the seaweed polysaccharides, carrageenan, which is found in red algae, can counteract cervical cancer. This type of cancer, which annually claims the lives of 250,000 women around the world, is caused by a virus that is transmitted sexually in the same manner as HIV and herpes. Carrageenan seems to protect the surface of the cells, preventing the virus from entering them. The surprising aspect is that carrageenan is more effective than the medicines currently on the market. With this discovery, both a preventative measure and a promising treatment are in sight. In addition, because the treatment would be inexpensive, it would be economically feasible to make it accessible to the poor. A large-scale clinical trial being conducted in South Africa and Thailand is trying to determine whether contraceptive gels that contain carrageenan will live up to expectations that they will also protect women against HIV infection. For the same reason, a company is producing condoms with a carrageenan coating.

Japanese women resident in Japan generally have a low incidence of breast and thyroid cancer. For example, the incidence of breast cancer in women is eight times lower in Japan than in Great Britain. On the other hand, Japanese women who live abroad, follow a Western diet, and eat little seaweed run the same risk of having breast cancer as women in the Western world. It has been suggested that the positive effect of seaweeds is due to their enhancing the metabolic breakdown of estrogen, especially estradiol, in the stomach and the intestines. Clinical studies have shown that by eating five grams of brown algae (*Alaria esculenta*) daily, post-menopausal women in the West can reduce the estradiol level in the blood to the same level as that of Japanese women.

Some researchers think that there is a connection between iodine and thyroid function on the one hand and breast cancer on the other. Iodine deficiency can be symptomatic of a risk of breast cancer and it is known that breast cancer cells have a lower iodine count than healthy breast cells. Nevertheless, it is obvious that factors other than seaweed consumption can be responsible for the lower incidence of breast cancer among Japanese women.

ARE THERE POTENTIALLY HAZARDOUS COMPOUNDS IN SEAWEEDS?

It is true that seaweeds can accumulate toxic compounds from the water or themselves produce secondary metabolites that may be toxic. Due to the diversity of seaweed species, however, it is impossible to make general statements about the level and type of potentially hazardous compounds in seaweeds. Moreover, the extent to which this may happen will depend on environmental conditions, time of harvest, and modes of processing. Special concerns pertain to heavy metals, iodine, inorganic arsenic, vitamin K, and neurotoxic amino acids. To illustrate why eating marine algae obtained from reputable sources should not pose problems, these substances are discussed below using dulse (*Palmaria palmata*) as an example.

Contamination from heavy metals is potentially hazardous in all food from the ocean, including seaweeds. Seaweeds absorb rapidly, and very effectively, cadmium, lead, copper, nickel, and mercury. In particular, mercury in the form of methyl mercury is of major concern since it can adversely affect the central nervous system, the endocrine system, and organs like the kidneys. Seaweeds do not actually require mercury for their metabolic functions, so the contents vary significantly depending on the pollution level of the ambient seawater. Normally, the levels of heavy metals in commercially available seaweeds are way below the safety limits imposed by regulatory authorities.

The possible nutritional importance of arsenic and the details of its metabolic function are unknown. It exists in both inorganic and organic forms. Whereas authorities do not always distinguish between these forms and refer to total arsenic when issuing recommendations, it is important to stress that it is arsenic in the inorganic forms, which is associated with the largest potential health risk. Upon ingestion, inorganic arsenic is quickly methylated in the liver and most of it is excreted with the urine. Inorganic arsenic has been used infamously, throughout the ages, to poison humans and animals. But in contrast to inorganic arsenic, there is no known adverse effect of organic arsenic. The average person's total intake of arsenic is about 10–50 μg/day. Values of about 1,000 μg are, however, not unusual following consumption of fish or mushrooms. Authorities have set no daily upper limit for the intake of arsenic from food. Since a major daily source of arsenic in some areas can be drinking water, some agencies have set a maximum upper limit of arsenic in water at 50 μg/L or even lower than 10 μg/L. As an example, total arsenic in dulse has been measured to be 1–10 μg/g. A portion of 5 grams of dulse will, therefore, maximally contain 5–10 μg arsenic, which is likely to be of no health concern even if a part of it is in inorganic form. The levels of inorganic arsenic in dulse are typically one to two orders of magnitude less that the total content of arsenic. The inorganic arsenic content depends on the age of the seaweed. For example, in some cases levels are found to be below the detection limit (0.02 μg/g) for young dulse, whereas older specimens can contain up to 0.3 μg/g.

Whereas arsenic is less of a concern in red algae like dulse, it can be a major problem in some brown seaweeds. In particular *hijiki* (*Sargassum fusiforme*), which is a staple ingredient in the Japanese cuisine, can contain very large amounts of arsenic. Some studies have found 141 μg/g, of which 85 μg/g is inorganic. Although a large part of this can be extracted in water during preparation, it is forbidden to sell *hijiki* as food in some countries.

Some brown seaweeds can contain very large amounts of iodine. The recommended daily intake of iodine is ca. 150 μg, but a healthy person can actually eat far more, for example, up to several milligrams. Healthy individuals will only experience difficulties from eating iodine-rich seaweeds when they are ingested in great quantities, but those who already suffer from thyroid problems should avoid eating too many seaweeds, especially kelp, which has a high iodine content. Too much iodine in the food can, in rare instances, cause medical problems in that it can trigger the thyroid gland either to over-produce or underproduce the iodated hormones that control metabolism. A few instances have been found in Japan where illness was caused by large quantities of iodine in the body, linked to an excess of seaweeds in the diet. In these cases, the build-up of iodine was attributed to taking seaweed-based

Allergies to seaweeds are rare. Furthermore, allergies to fish and shellfish have no correlation with a possible intolerance of seaweeds.

food supplements in the form of capsules of dried and granulated kelp, which contains more iodine than other seaweeds.

Vitamin K is an umbrella term for a group of fat soluble compounds that come in two natural variants, vitamin K_1 and vitamin K_2. Vitamin K_1 is synthesized in green plants and algae and K_2 in bacteria, e.g., in the human colon. None of the vitamin K types are known to be toxic as such to humans, even in large oral doses of 20mg or more, and no adverse effects have been reported from human consumption of vitamin K in food or supplements. But since vitamin K is a co-enzyme required for the formation of blood-coagulation factors, excessive and uncontrolled intake of vitamin K, for example, from seaweeds in the diet, may interfere with medication in individuals who are taking blood-thinners, e.g., warfarin. Clinical studies have shown that intermittent and smaller changes in vitamin K intake do not, however, require permanent changes in the dosage of warfarin.

The adequate daily intake of vitamin K for adults is 90–120μg and recommended daily intakes are 200–500μg. Large variations in dietary vitamin K_1 are known to be caused by eating fresh parsley (1,640μg/100g), spinach (483μg/100g), water cress (250μg/100g), and broccoli (103μg/100g). Measurements show that dried dulse may contain 200–700μg/100g vitamin K_1. Since a typical daily intake of dried dulse is maximally about 5 grams it appears that even individuals taking a blood-thinning medication need not worry about a diet containing moderate amounts of dulse.

A special concern has been raised regarding the possible presence of domoic and kainic acids in some red seaweeds. Kainic acid is an amino acid belonging to a group of compounds known as kainoids. These compounds are neuroactive, and the most potent kanoid is the amino acid domoic acid that can cause amnesic shellfish poisoning. Whereas domoic acid has been found in some red macroalgae, it is predominantly produced in diatoms that can accumulate in shellfish and other marine organisms. Kainic acid has been found in a few different macroalgae, e.g., in some strains of dulse where it is a secondary metabolite. Research has shown that some strains of dulse are producers of kainic acid, whereas others are non-producers. Recent analyses of kainic acid in dulse from several different locations and sources show variations from 1–130μg/g. Animal studies with rats and mice have demonstrated that injection of kainic acid into the brain or body cavity can lead to seizures, hippocampal damage, and behavioral changes. The doses used are, however, extremely large, typically from 4–32mg/kg body weight.

Doses of those orders would correspond to up to 2 grams pure kainic acid for an adult human being. There appears to be no published data regarding human safety values, neither are there there any published studies relating oral intake of food containing kainic acid to neuronal activity in humans. In order to reach the hazardous levels of kainic acid used in the mice and rat experiments, a total amount of about 30 kilograms dried dulse of the variety with the highest concentration of kainic acid is required. It is highly unlikely that a human being would consume such a large amount of dulse in one meal. Furthermore, the consumed dulse has to pass through the gastrointestinal system before possibly making it into the bloodstream and from there across the blood-brain barrier. It would appear, therefore, that consumption of most dulse species does not present any serious danger to human health.

HOW MUCH SEAWEED SHOULD ONE EAT?

There is no scientific evidence for dietary advice about how much or how little seaweed one should eat. As there is also no well-documented proof of the therapeutic effects of marine algae, there are no recommended dosages for seaweed products in a medicinal context.

Nevertheless, there are two relationships that need to be taken into account. One is the iodine content in seaweeds and the other is the content of salts. In both cases, the central issue is to eat neither too much nor too little. Not surprisingly, the well-known maxim of the Renaissance doctor and alchemist Paracelsus still applies: All substances are poisons; it is only the proper dose that differentiates a poison from a remedy.

Even though seaweeds contain potassium salts, and in the case of some species a greater quantity of potassium salt than of sodium salt, the overall salt content can pose a problem for persons with a tendency to high blood pressure. So the golden rule is the usual one: use salt sparingly.

In the normal Japanese diet, which typically includes about 4–10 grams of marine algae every day, the iodine content is about 1 mg. It has been estimated that the average daily Japanese iodine intake is 1–3mg and that some Japanese consume as much as 20mg of iodine in their daily food. In a single sheet of *nori*, used for sushi rolls, there are only about 40µg of iodine, but it is not unusual for a bowl of good *miso* soup made with *konbu* to have about 1mg.

Our intestinal flora and enzymes are not normally adapted for breaking down some of the carbohydrates found in seaweeds. Consequently, some people benefit from incorporating marine algae into their diet slowly to allow their systems to adjust. It is, therefore, sensible to eat small portions of seaweeds regularly, instead of big quantities once in a while. Also, one should not expect that any positive effects of seaweeds on overall health would be noticeable immediately; this can take several months.

Given all of the above, it seems reasonable to recommend that the average adult could eat about 5–10 grams dry weight of seaweeds on a daily basis.

Seaweeds—the hidden treasures in the Natural History Museum in London

The Natural History Museum is located in South Kensington, a fashionable area of London, just a short walk from Harrods department store. One approaches the building, which is set above street level and is accessed by an imposing staircase leading to the monumental entrace doors, with a sense of humility. Inside the cathedral-like Central Hall, which was completed in 1880, one meets the museum's icon, a copy of the skeleton of a 26 meter long dinosaur, *Diplodus carnegii*, known affectionately as Dippy. It is one of the ca. 70,000,000 artifacts and specimens housed in the museum, the largest of its kind in the world.

Somewhat naively, I had gone to see what the museum had by way of seaweed specimens in the botanical collections. It quickly turned out, to my great disappointment, that there were none on display. The pleasant young woman at the information desk could tell me only that I had to contact the botanical department of the museum and passed on a telephone number.

Back in my room at the Royal Society with a view of St. James Park and Big Ben, where I was lucky enough to be lodged for a few days, I immediately started calling and managed to get the algae curator herself, Jenny Bryant, on the phone. I would be allowed to see the collections the next day and she would show me around. When Jenny Bryant met me in the front hall of the museum, she took me up to the top story and through an undistinguished door with a sign over it that reads *Cryptogamic Herbarium*. On the other side of the locked door is a treasure trove to which the general public has no access. This is an enormous room with cabinet after cabinet where the algal collections are neatly ordered behind beautiful brown wooden doors that protect the fragile specimens from exposure to light.

I was shown examples of the collections' stock of green, red, and brown macroalgae. There are about 600,000 seaweed specimens, with the oldest dating from the 1700's. To my great surprise and delight, the first folio that Jenny Bryant pulled out was from the personal collection of Kathleen Mary Drew-Baker, the famous researcher who unlocked the secrets of the *Porphyra* life cycle, paving the way for the modern Japanese *nori* industry. Here I could see with my own eyes her carefully mounted *Porphyra umbilicalis* and read annotations, in her own handwriting, about classification and where the specimen was found.

Then I was allowed to sit amongst the collections and make use of the library, where there are many interesting books about algae that I have found nowhere else. Sitting beside the old shelves and cupboards, I soaked up the congenial atmosphere of this place, where time seemed to stand still and the staff went quietly about their tasks.

When I was leaving, Jenny Bryant told me that it had been a good day for algae, as one of their former colleagues had dropped by. I had the impression that seaweed research is a low priority at the museum and that only a few phycologists work on the collections. I exited by the little side door and found myself back on the large main staircase. To think that one could pass by that modest entrance and have no idea about what lies behind it.

'Seaweed people' are friendly and they are possessed of a peculiar, quiet enthusiasm for that special alga to which much of their working life is devoted. And they love to share their passion for it with others who are interested in the subject.

I left the museum in a good mood, with my own seaweed obsession recharged. Better yet, an arrangement had been made for me to come back again and have photographs taken of some selected specimens from the collections.

▶ Dried specimen of the red alga sea beech (*Delesseria sanguinea*) from the collections of the Natural History Museum in London.

▶▶ The Natural History Museum in London and a peek into the Herbarium.

114

Seaweeds in the kitchen

Seaweeds in the home kitchen

SEAWEEDS ARE EATEN ALL OVER THE WORLD

Although brown, red, and green seaweed species have been eaten by all coastal peoples since prehistoric times, their regular consumption has survived to the present time, primarily in the modern cuisines of Asian countries such as Japan, Korea, China, and the Philippines.

In Japan, especially, seaweeds are both fully integrated into the daily diet and used to create gourmet specialties. Everywhere in East Asia, there is a wealth of well-known traditional dishes incorporating seaweeds that reflect the great diversity of national and regional cuisines. An impressive variety of species, as well as blue-green microalgae, are used in a wide range of foods,

▶ Fresh salted *wakame* for sale at a street market in Katsuura, Japan.

such as soups, salads, desserts, pickles, snacks, and flavorings. Elsewhere, they are overwhelmingly associated with the preparation of Asian-style meals, particularly sushi, which has become a global phenomenon.

Nevertheless, while the practice of eating seaweeds as part of the normal diet has almost died out in the Western world, it can still be encountered here and there. For example, within Europe, seaweeds are firmly entrenched in the popular cuisines of Brittany, Wales, and Ireland. On Iceland, dulse is eaten in dried form as a snack or mixed into salads, bread dough, or curds, just as it was in the time of the sagas. And the pattern of consumption is on

118

an upward trend, no doubt driven by an awareness of the beneficial effects of seaweeds on overall health.

In Europe, and particularly in Ireland, there are encouraging signs of a general revival of interest in cooking with seaweeds, even though this is still considered a bit quirky and exotic. In North America, eating seaweeds is also gaining in popularity, especially in California and Maine in the United States and British Columbia and Nova Scotia in Canada, where selected dried seaweeds are available in many supermarkets. Internet-based small businesses are emerging around world, taking advantage of the easy shipping of dried goods.

At the level of gastronomy, marine algae are also making inroads. Many of today's leading chefs have become fascinated with seaweeds and are experimenting with their tastes, textures, and colors in familiar dishes and in exciting new gastronomic creations. In many cases, this involves finding interesting ways to come up with recipes that fuse several cuisines or that revitalize national dishes and the food cultures of former times.

Seaweeds in the home kitchen

SEAWEED RECIPES

Probably no other ingredients found in the kitchen are as versatile as seaweeds. They can be eaten raw, cooked, baked, toasted, puréed, dried, granulated, or deep fried. They can be eaten on their own or combined in countless ways with other cold or hot ingredients. In most instances, the seaweeds retain the greater part of their nutritional value in unchanged form. Since seaweeds are easy to store when dried, it takes little effort to keep them at hand in the home kitchen.

At first, most people probably find the taste and texture of seaweed dishes a bit strange. For this reason, it is a good idea to accustom oneself to eating algae gradually, for example, by adding them in small quantities to dishes that are already favorites. Dulse is a good starting point, from which one can slowly move toward the milder varieties such as *wakame* and winged kelp, before eventually venturing on to *nori* and *hijiki*. To derive maximum benefit from the range of vitamins and minerals found in seaweeds, it is best to incorporate a variety of brown, green, and red algae into one's diet.

Most of the recipes in this chapter are uncomplicated. Some are designed to demonstrate how easy it is to enhance the nutritional value and taste of a range of familiar dishes just by adding a little seaweed to them. A few of the others are there in the hope that the reader who has not already done so may be tempted to try seaweed for its own sake.

It is unfortunate that marine algae came to be known in English as 'seaweeds'—many people find the idea of eating 'weeds' off-putting. However, as more is known about how edible they are, attitudes are slowly changing. In her classic seaweed cookbook published in 1977, Judith Cooper Madlener states emphatically that they should be called 'sea vegetables'. Others have introduced terms with positive connotations, such as 'vegetables from the sea' and 'sea greens', which conjure up images of foods that are healthy and nutritious.

STORAGE OF SEAWEEDS

It is virtually impossible to find fresh seaweeds in ordinary shops and very few people have the opportunity to gather them at the seaside for their own consumption. Even in Japan, only a few species of seaweeds are sold fresh, the most common being salted, newly harvested *wakame*. Proper conservation and storage of seaweeds are, therefore, both necessary to ensure a steady supply of these ingredients.

Fresh algae start to break down and change color very rapidly after they are harvested; in fact, they do so considerably more rapidly than terrestrial plants. Most probably, this is due to differences in their cell walls, which in seaweeds are more permeable, thereby speeding up the decomposition of those pigments that are responsible for their characteristic colors. In addition, bacteria are often present on the surfaces of the blades. Hence, fresh seaweeds must be processed right away. Normally, they are first washed in clean seawater or a saline solution, dried either in the shade or in direct sunlight, and then stored away from light in a dry environment. Some types of brown algae benefit from being dried in sunlight, as ultraviolet rays help to convert the seaweeds' polyphenols to simple tannins, which are an important element in how they taste.

In dried format, edible seaweeds are most easily stored either as whole blades or as granules. Typically, dehydration takes place at around 40°C, a

▸ An assortment of dried Japanese seaweed products.

temperature at which the color and vitamin content suffer the least damage. If the dried seaweeds are stored in an airtight package or in a glass with a tight-fitting lid, they can keep for years. Their taste and color do, however, change over time, especially if the seaweeds are exposed to humidity or light. In general, it is also true that preserved algae quickly lose a significant proportion of their vitamin content, especially vitamin C.

Dried *nori* (*hoshi-nori*) has a very dark greenish-black color, but after it has been stored for a long while it turns violet or rose-colored and loses some of its flavor. The change in color is due to enzymatic breakdown of the green pigment, chlorophyll *a*. Some *nori* manufacturers, therefore, destroy these enzymes by heating the seaweeds before packaging them. *Nori* is not well suited for freezing as is exudes a red pigment as it thaws. It also absorbs moisture readily and loses its keeping qualities once it is damp. Nevertheless, a slightly moist sheet of *nori* can easily be dried and made crisp again by warming it for a few seconds in a toaster oven or holding it over a gas flame.

As their cell walls are stronger, brown algae are more robust and they are, therefore, easier to store than red algae. They seldom change color because their pigments are more numerous and more stable and it is not necessary to store them in completely airtight packages. In addition, they naturally contain alginates and fucoidan that protect them against attack by bacteria and fungi. Sometimes a layer that resembles white dust forms on the surface of dried *konbu* and dulse after they have been stored for a period of time. Along with other substances, this layer contains MSG, a naturally occurring substance that imparts the *umami* taste. For this reason, these algae should not be rinsed before use in such dishes as soups. If *konbu* has faded and its dark green color has turned pale and yellowish, it is a sign that it is too old and has lost much of its nutritional value.

Dried *arame* is jet-black and is probably the seaweed species that keeps best when dried; in this way it can be preserved for years in an airtight package.

Normally, dried seaweeds are rehydrated in water before use, for a period that varies from one minute to a couple of hours, depending on the species. It is recommended that water should be used sparingly, so that as little as possible of the seaweeds' mineral content is drawn out. In the course of this process, the volume of the algae typically increases to several times that of the dried original. Surplus water is removed by squeezing the seaweeds lightly with one's hands. Once rehydrated and softened, the seaweeds can keep in the refrigerator for about a day.

Seaweeds in the home kitchen

▲ *Wakame* seaweed salad (*hiyashi wakame*), which is often commercially available. Some variations of this product also contain fungi, such as Jew's ear (*Auricularia auricula-judae*), which adds a more crunchy texture.

Seaweeds can also be preserved by salting or canning them in the same way as ordinary vegetables. In France, fresh dulse, thongweed, and winged kelp are salted and can keep for 4–12 months in a cool place. Before use, they are soaked in a large quantity of water to remove some of the salt and agitated lightly to rinse off any small shellfish that may have attached themselves to the blades. Excess water is then squeezed out by pressing them gently with one's hands. As with rehydrated algae, the desalted seaweeds can then be kept in the refrigerator for about a day.

Seaweeds as seasonings

SEAWEEDS IN THE SPICE RACK

Seaweeds can be used with great success both as salt substitutes and as flavor enhancers. An easy, direct way to take advantage of these wonderful properties is to use seaweed powder or granules, which are usually made from a blend of *Laminaria* species or bladder wrack. The powder can be shaken, like salt, on warm, cooked dishes or on prepared foods such as salads.

Several things can be gained by substituting seaweed powder or granules for salt. In the first instance, a teaspoonful of seaweed powder is sufficient to cover our daily iodine needs and it is also a good source of calcium, iron, and vitamins B_1, B_2, and B_{12}. Secondly, seaweeds contain a great deal of potassium, often more potassium than sodium. Seaweed can replace salt in food, e.g., in bread. This means that one can enjoy the taste of salt without running the risk of raising one's blood pressure, a decided advantage for those who have a tendency toward, or suffer from, hypertension. Thirdly, seaweeds add nuances of flavor to the food, not the least of which is *umami*.

▶ Seaweed powders and granules produced from bladder wrack and winged kelp that have been cleaned, sun dried, and crushed.

FURIKAKE—A JAPANESE CONDIMENT WITH SEAWEEDS

A whole arsenal of seasonings based on seaweeds is to be found in the Japanese kitchen. In every well-stocked Japanese food store or supermarket, one can find a wide assortment of these blends, known as *furikake*, which are often sold in shaker jars or in serving-size packages. Many of the mixtures contain dried and toasted algae, in the form of either small granules or thin strips. They are often combined with toasted black and white sesame seeds, dried fish flakes (*katsuobushi*), garlic, cayenne, or chili pepper. Although toasted *nori* makes up the usual seaweed component of *furikake*, dried, granulated flakes of green algae such as *Ulva* or *Monostroma*, known as *ao-nori*, can also be used.

Furikake imparts *umami*, savory, salty, and spicy tastes to foods. The most common way of using this seasoning is to sprinkle it over warm, cooked rice. The dried seaweed bits give off a fantastic aroma when they are softened by contact with the warm moisture of the rice. Similarly, *furikake* can be used as a topping on steamed fish, vegetables, and salads.

A word of advice—because dried seaweeds, especially *nori*, quickly absorb moisture, they should be added only when the meal is ready to be served to have maximum enjoyment of the crispness of the *furikake* and the aromas that are released.

I use *furikake* in my cooking virtually every day—just a little bit adds subtle colors and flavors to an otherwise ordinary dish. It is a staple you want to have readily available on your lunch and dinner table. And I have to confess that I often travel with *furikake* tucked away in my suitcase. On occasion I have been taken aside in an airport to declare it to customs as a 'plant-based' foodstuff, which it most certainly is not—algae are not plants!

▲ *Furikake*—a Japanese condiment with *nori*, often sprinkled on prepared food.

SEASONING WITH SEAWEEDS

Just about anyone, anywhere, can make a simple, versatile version of *furikake*. Simply toast dried pieces of winged kelp, *nori*, or dulse just before use and then crush them by hand and sprinkle the bits on the food. A small packet with miniature sheets of toasted *nori* (*yaki-nori*) to be eaten with warm rice is traditionally served as part of a Japanese breakfast.

Apart from using crushed seaweeds in the conventional ways with fish and vegetables, they can even be combined with fruits. When I visited Louis Druehl and his wife Rae Hopkins on Vancouver Island, I was surprised when they served fresh, cut up fruits such as apples, oranges, and bananas with a sprinkling of crushed and toasted winged kelp. But its aromatic and slightly salty taste was a perfect complement for the sweetness of the fruit.

▲ Granulated dulse.

Seaweeds in soups

Seaweeds in the kitchen

Different types of seaweeds, especially the brown algae, work well in soups. *Konbu* (*Saccharina japonica*), which is coarser, is very useful for making stock, while the more delicate *wakame* and winged kelp can be eaten in the soup. *Konbu* contributes *umami* taste and salt to the soup, while *wakame* adds color and an interesting texture. Soups with a seaweed base are a staple of Asian cuisines, particularly that of Japan.

JAPANESE SOUPS AND *DASHI*

Most Japanese soups are based on a stock, *dashi*, made with *konbu* and dried fish flakes. The fish flakes, *katsuobushi*, are produced from a fish popularly referred to as 'bonito', which is actually skipjack tuna (*Katsuwonus pelamis*). Fillets from this fish are cooked, salted, smoked, fermented, and dried to form rock-hard pieces that are shaved into paper-thin flakes. This fivefold method of preserving the fish draws out an extraordinary range of taste and aroma substances, some of which resemble those in dried, salted ham and cheese. For its part, the *konbu* contributes soluble minerals and amino acids to the stock. As these have often seeped out onto the surface of the *konbu* blades, the seaweeds should not be washed before cooking.

The raw ingredients in *dashi* are used twice. When they are extracted in water the first time, the result is *ichiban dashi*, a delicate, clear broth, which is the finer of the two stocks. They are then boiled a second time to make *niban dashi*, a cloudy, more strongly flavored base used in soups and other dishes. For those in a hurry, there is another, faster way to make *dashi*. Stores carrying Japanese food products often sell *dashi* powder in small, practical packages that are sufficient to prepare one bowl of stock. This powder is flavored with both *konbu* and *katsuobushi* and often contains MSG.

First *dashi* (*ichiban dashi*)

To make this stock, place 10–20 grams of *konbu* in 1 liter of cold water and heat gently to below the boiling point. Avoid breaking the seaweed into smaller pieces as this would allow alginates to seep out of the blades, causing an undesirable thickening of the soup. The water must not be allowed to boil as this would cause the *konbu* to give off a bitter taste. In fact, experiments show that the best flavor of the first *dashi* is obtained when keeping the seaweed in water at 60°C. Remove

▶▶ Dried carrageen on top of dried *konbu*.

the seaweeds and add 10–20 grams of *katsuobushi* flakes—the greater quantity results in a more strongly flavored stock. Bring the mixture to the boiling point and then remove from the heat. Add about 100 milliliters of cold water and allow the stock to rest for a few minutes. Then pass it through a sieve, lined with cheesecloth, to separate out the clear broth. First *dashi* is used to make a clear soup, *suimono*, and is also the liquid in which *shabu-shabu*, a Japanese dish a little like meat fondue, is cooked.

Second *dashi* (*niban dashi*)

The *konbu* and the *katsuobushi* from the first *dashi* are recycled to make a second stock. Place them in 2 liters of water and bring it to the boiling point. Then let the stock simmer for about half an hour. Strain through a sieve to extract the fish flakes and the seaweed pieces. Second *dashi* is used to make *miso* soup and may be included in other dishes.

CLEAR SOUPS—*SUIMONO*

Suimono is the most minimalist soup imaginable. It is made from *dashi* and has only a few additional ingredients, but preferably ones with shapes and colors that complement each other.

Suimono

Heat a portion of first *dashi*, to which a few ingredients have been added. Examples are two slices of raw mushroom, e.g., *shiitake*, a couple of pieces of *wakame*, a single prawn, a piece of *tofu*, or a little piece of white fish with the skin on. Serve the *suimono* in a small soup bowl with a lid, which can be replaced between sips to keep the soup warm. Chopsticks or a spoon are used for fishing out the larger solid ingredients of the soup.

Seaweeds in the kitchen

Soup in Japanese *haiku*

nori jiru no
tegiwa mise keri
asagi wan

seaweed soup
shows such skills
in a decorated bowl

Matsuo Bashō (1644–1694)
(translated by Jane Reichhold)

▶ *Suimono*, a clear soup, with *wakame*.

▲ Dried *wakame*.

Miso soup

Miso soup is made from *dashi*, to which white, red, or dark *miso* paste is added. This soup is a good source of proteins and supplements the nutritional elements found in rice. Additional ingredients used in *miso* soup include seaweeds like *konbu* and *wakame*, *shiitake*, spring onions, small pieces of fish, and shellfish. Dried seaweeds and mushrooms must be rehydrated before use. As is the case for a clear broth, these additions should be few in number and variety. It is desirable to have one that floats and breaks up the surface of the soup. *Miso* is normally served in a bowl from which one can drink; the larger solid pieces are eaten with chopsticks or a spoon.

Seaweeds in soups

Miso soup

To make 1 liter of soup, use ca. 1 liter of second *dashi* and 2 tablespoons of *miso* paste. As *miso* paste must not be boiled, it should be put in at the end. Add the other ingredients to the soup in accordance with the cooking time they require. Some mushrooms and vegetables take longer to cook, whereas small pieces of fish and *tofu* essentially need only to be heated through. Dissolve the *miso* paste in a little lukewarm water and add it to the soup. Stir it in, heat the soup to just below the boiling point, and serve immediately.

St. Patrick's cabbage soup

The feast of St. Patrick, the patron saint of Ireland, is celebrated each year on the 17th of March. His name has been given to a traditional Irish cabbage soup, which was originally made with the help of the red alga carrageen (*Chondrus crispus*). This soup is a rather heavy vegetable soup, but is nourishing and just right on a cold day. In some classical recipes, dulse is used instead of carrageen.

St. Patrick's cabbage soup   *serves 4*

15g dried carrageen, cut into small pieces
400ml savoy cabbage, finely chopped
3 medium sized potatoes, peeled and diced
100g mushrooms, sliced
1 medium onion and 1 clove garlic, finely chopped
a few sprigs of parsley, coarsely chopped
1 liter chicken bouillon
25g butter

▲ Dried *konbu* (*Saccharina japonica*). The blade is about 1.5 meters long and folds in the middle.

127

Soak the seaweed in cold water. Melt the butter in a large pot, add the onion, garlic, and potatoes and fry lightly for 5 minutes. Add the cabbage and mushrooms. When all the ingredients are soft, add the chicken bouillon and the carrageen. Allow the soup to simmer for 10 minutes and season to taste with salt, pepper, and chopped parsley.

NEREO SOUP

This recipe was given to me by Rae Hopkins, who maintains that this tasty soup can be prepared in the wink of an eye. It is based on the brown alga bullwhip kelp, *Nereocystis luetkeana*, for which it is named. If dried bullwhip kelp is not available, *wakame* or winged kelp can be substituted.

▲ Fresh carrageen.

Nereo soup	*serves 4*

15g dried bullwhip kelp, cut into small pieces
125g *tofu*, diced
1 liter chicken or vegetable bouillon
125g broccoli florets
80ml soy sauce
1 egg, beaten lightly
1 or 2 spring onions, finely chopped

Warm the bouillon together with the seaweed, *tofu*, broccoli and soy sauce. Stir the beaten egg into the soup. Serve with a sprinkling of spring onions.

DILLISK SOUP—A SIMPLE SOUP WITH DULSE

Ireland has a rich tradition of using algae in soups, both as thickeners (for example, carrageen) and as main ingredients. *Dillisk* is the Gaelic name for dulse. In earlier times, fresh dulse was gathered at ebb tide and spread out to dry in the sun on a tin roof or on roofing stones. The dried dulse could be eaten raw between two pieces of buttered bread or used in a stew or soup.

Dillisk soup	*serves 4*

10g dried dulse, or toasted purple laver, cut into slivers
2 liters water
15g green lentils

1 stalk celery, chopped
2 small potatoes, peeled and chopped
1 medium onion, chopped
oil, lemon juice, cayenne, herbs, spring onions

Sauté the dulse, celery, and onion in a little oil for about 5 minutes.
Add the potatoes and lentils, together with the water. Boil the soup
for about 20 minutes, possibly with some fresh herbs and a little
cayenne. Blend the soup, season to taste with lemon juice, salt, and
pepper. Serve with a sprinkling of crushed, toasted dulse and a few
slices of spring onion.

JERUSALEM ARTICHOKE SOUP WITH SEAWEEDS AND SMOKED SEA SALT
Jerusalem artichokes (also called sunchokes) are related to sunflowers and
have edible tubers. Like onions and garlic, they are lacking in starches and
instead use the carbohydrate inulin as an energy store. Inulin cannot be
broken down by human digestive enzymes, which are adapted for converting
starches. As Jerusalem artichokes consequently have no effect on blood-sugar
levels and contain fewer calories than potatoes, they are a good food source
for diabetics. In addition, the plant fibers in these tubers are soluble, leading
to gelatinization and water absorption in the intestines, which helps to lower
cholesterol and glucose levels in the bloodstream. But one unfortunate side
effect is that some may find them difficult to digest. Even though inulin cannot
be converted by the digestive enzymes, it can be broken down by bacterial
flora in the colon and thereby produce carbon dioxide or methane, which
causes bloating and flatulence.

Jerusalem artichokes have a crisp texture and, when raw, a delicate, nut-
like taste, which becomes a little sweeter but a bit insipid when the tuber is
cooked. Smoked sea salt and seaweed granules are, therefore, a perfect way
to add smoke, salt, and *umami* flavors.

Jerusalem artichoke soup with seaweeds *serves 4*
1 tbsp granulated seaweeds (*wakame*, bullwhip kelp, or giant kelp)
500g Jerusalem artichokes
100ml milk or cream
2 tsp aromatic, vegetable sea salt (e.g., Herbamare)
water

1 medium onion or spring onions
1 clove garlic
smoked salt
pepper

Rinse and thoroughly scrub the Jerusalem artichokes or peel them if they are very gnarly and difficult to clean. Slice the tubers, put them in a pot with just enough water to cover them, and cook with onion and garlic together with the aromatic, vegetable sea salt for 10 minutes. The slices should remain somewhat crisp and must not be allowed to overcook. Then blend it all, leaving a few small chunks in the mixture, and return to the pot with the water and the granulated seaweeds. Warm the soup, add the milk or cream, and adjust the seasoning with pepper and smoked salt. A few extra seaweed granules can be sprinkled on top at the time of serving.

▲ Smoked salt.

Green pea soup with scallops and seaweeds

A wonderful summer soup can be made from green peas, flavored with tastes of salt, smoke, and seaweeds. By adding scallops to this soup you will obtain an intense *umami* taste produced by the combination of peas, scallops, and seaweeds.

Green pea soup with scallops and seaweeds   *serves 4*

5–10g dried kelp, winged kelp, *wakame*, or dulse, rehydrated in water
500g green peas, fresh or frozen
4 large or 8 small scallops
1 medium onion, in chunks
1 clove garlic
2 tsp aromatic, vegetable sea salt (e.g., Herbamare)
water
granulated kelp
pepper and smoked salt

Bring the peas to a boil in just enough water to cover them. Blend the seaweeds with a little of the cooking water, leaving them in coarse pieces. Add the cooked peas, onion, garlic, and the rest of the cooking water and blend very briefly. If necessary, add water to achieve the desired consistency. The peas should not be completely liquefied, so as to encourage a little chewing when the soup is served. Add smoked salt, pepper, and possibly some granulated kelp to adjust the seasonings. Keep the soup warm. Sear the scallops lightly on both sides on a hot, dry frying pan. Just before serving, ladle the soup into bowls, add the scallops, and sprinkle a few flakes of dried seaweeds on top.

◀ Green pea soup with scallops and seaweeds.

KONBU CONSOMMÉ WITH SHIITAKE

This recipe for a simple soup with brown algae and fungi serves up a veritable orgy of *umami* tastes. It is filling even in small portions.

Konbu consommé with *shiitake*	serves 4

20g dried *konbu*, winged kelp, or sea tangle

2 dried *shiitake*

1 small piece of fresh ginger

½ liter water

2 fish or vegetable bouillon cubes

soy sauce or *miso* paste

thyme or basil

Soak the *shiitake* until they are soft. Also soak the seaweeds, which can take up to about an hour if *konbu* is used. Discard the soaking water, drain the fungi and the seaweeds, and cut them into small pieces. Peel the ginger and slice thinly. Cook all the ingredients for about 15 minutes in the water to which the bouillon cubes have been added. Adjust the seasonings with soy sauce or *miso* paste and possibly a little fresh thyme or basil.

Seaweeds—a gift from the sea to the first inhabitants of the Pacific Northwest

The traditional culture of the First Nations peoples and Native Americans, who were the original inhabitants of the Pacific Northwest coast of North America, is closely connected, both physically and spiritually, to the sea and its animal and plant life. Seaweeds, especially red and brown algae, were an important part of the food that the sea gave to humankind.

Red algae of the *Porphyra* genus (*Porphyra abbottiae*) were harvested by many different indigenous peoples and continue to be popular among some of them to this day. The seaweeds were used both as a local food source and as a trading commodity that was exchanged with those who lived inland. At the beginning of the 1900's, First Nations peoples living around Vancouver in British Columbia gathered *Porphyra* and sold it to Chinese and Japanese who had settled in the area. It is still harvested by them commercially and is eaten to prevent goitre.

Traditionally, the red algae were harvested at low tide, especially in the month of May, while they were still tender. They were then treated in different ways to preserve them for consumption right through the winter. The simplest method was to dry the seaweed blades on a warm stone in the sun. Once dried, they were broken up into small pieces and stored in a wooden box. The pieces were eaten as is, or cooked with fat, fish heads, or clams, mussels, or oysters. According to folklore, this type of food was good for the digestion.

Another preservation method was to suspend the seaweed blades over an open fire, so that they were both roasted and smoked. Once dried, they were crushed to a powder, which was easy to store. When it came time to eat it, the powder was mixed with water and cooked or whipped up into a foam and served as a sort of dessert.

A more elaborate way of treating and preserving the seaweeds seems also to have involved adding other taste elements to them. The fresh *Porphyra* blades were dried off lightly and fermented, then placed in a tall cedar chest in layers. Cedar planks were placed between the layers to impart

◂ Mythological depiction of bullwhip kelp (*Nereocystis luetkeana*) by G̲iitsx̲aa, a Haida artist from the Pacific coast of British Columbia.

their flavor and sometimes liquid from bivalves was poured onto them. Lastly, a stone was placed on top to compress the layers. After a month or a little longer, the finished product was ready—a flat seaweed cake, a few centimeters thick, which could be kept over the winter. Preparation consisted of chopping the cake into pieces with a sharp adze and then cooking them in water together with a little fat. The result was a sort of soup, which was often eaten with cooked halibut heads or mussels, clams and oysters. Etiquette dictates that at feasts one should drink water only after eating this dish, not before.

The majestic appearance of *Nereocystis* in the ocean has also given rise to legends and myths. According to a wonderful legend told by the people of Haida Gwaii in British Columbia, if one encounters a *Nereocystis* with two heads, that is to say, two air bladders, one will come face to face with a powerful, supernatural creature from the sea.

Seaweeds in salads and sauces

Green, red, and brown algae can all be used to make salads. Some types of freshly harvested seaweeds—for example, sea lettuce, dulse, or young blades of sugar kelp—can be eaten completely raw, but generally speaking most people find fresh seaweeds tough and somewhat unpalatable. One solution is to marinate the seaweeds, which helps to soften them and make them less perishable so that they can be kept in the refrigerator for several days. Marinated seaweeds can be served as a condiment, an appetizer, or mixed into a green salad.

A variety of seaweeds can be used to add a spectrum of colors and taste nuances to a simple green salad. In addition, their crisp textures can pep up an otherwise slightly dull salad that lacks crunch. The easiest ways to incorporate seaweeds into a green salad are to crush toasted *nori*, dulse, or *wakame* and sprinkle them over the salad leaves or to mix in thin strips of cooked and marinated algae. The question arises about how much seaweed should be added to a serving of salad for an adult? Normally it would be about 4–5 grams of dried seaweeds, which corresponds to about 20–30 grams when rehydrated. *Konbu* can act as a laxative, so one should proceed cautiously, starting out with smaller quantities.

Kaiso-salad

Kaiso is a generic Japanese term for all varieties of edible seaweeds. So *kaiso*-salad is simply a mixture of one or more types of seaweeds, tossed with a light dressing. It goes especially well with grilled or cooked fish, vegetables, shellfish, and smoked meats.

In Asian food stores and at fishmongers, one can sometimes buy ready-made *wakame* salad (*hiyashi wakame*), which consists of cooked, finely julienned *wakame* tossed with a little sesame oil and seasoned with chilies and sesame seeds. It is usually sold frozen and then defrosted just before use, but it can easily keep in the refrigerator for one or two weeks after it has been thawed.

A colorful *kaiso*-salad can be prepared very quickly from one of the mixtures of dried seaweeds that are sold in airtight packages, which typically include brown, green, and red varieties. The salad is ready to eat after the algae have been soaked in water for a few minutes, drained, and then dressed with rice vinegar, soy sauce, sesame oil, and possibly a few crushed chili pepper flakes. Toasted white or black sesame seeds are often sprinkled on top.

▲ The delicate Japanese red alga *funori* (*Gloiopeltis* spp.), which is delicious in a seaweed salad.

▲ *Ogonori* (*Gracilaria* sp.), a common red seaweed, some species of which are now considered invasive in large parts of Europe.

▲ *Tosaka-nori* (*Meristotheca papulosa*), a very fine red seaweed species that comes in three different colors – white, green, and red. As it is both crisp and colorful, it is very useful in salads and for decoration.

Seaweeds in salads and sauces

◄ *Kaiso*-salad—mixtures of different varieties of seaweeds.

Mixtures that contain a lot of *wakame* can look a bit as if they have been squashed because the thin blades of this type of seaweed tend to stick together after they have been rehydrated. One way to counteract this is to add ingredients that are firmer, crisper, or leafier, such as lettuce, green asparagus, purslane, or avocado. Another possibility presents itself if one is lucky enough to be able to obtain the delicate Japanese red alga *funori* (*Gloiopeltis* spp.). When soaked in water, it spreads out and forms a bushy branch.

Other red, green, and brown seaweeds are served as salads or appertizers, e.g., *mozuku* (*Cladosiphon okamuranus*), *ogonori* (*Gracilaria* sp.) and *tosaka-nori* (*Meristotheca papulosa*). Served in combination, they make a colorful dish that is pleasing to both the eye and the palate.

SALAD DRESSING WITH *MISO* AND SEA LETTUCE

Crushed seaweed bits or seaweed flakes, especially if made from winged kelp, dulse, or sea lettuce, can replace the herbs that are added to many dressings, regardless of whether they are based on an oil-vinegar mixture or made with plain yogurt or *crème fraîche*. For a salad prepared from cold, cooked green beans or fresh sugar peas, I prefer a simple dressing that uses white *miso*. It is also good on other green salads, such as one made with purslane.

▲ *Mozuku* (*Cladosiphon oka-muranus*) is a brown alga grown around the islands of Okinawa in Japan. It is typically cooked and marinated in salt and vinegar and served cold as an appertizer, a salad, or with *sashimi*. *Mozuku* has a soft and viscous texture.

135

Salad dressing with *miso* and sea lettuce

1/2 tbsp sea lettuce flakes

2 tbsp white *miso* paste

3 tbsp water

Mix the *miso* paste with the water and stir in the sea lettuce flakes.

PURSLANE SALAD WITH SEAWEEDS

Purslane is a shamefully neglected vegetable in the modern kitchen even though it can be used to prepare tasty, healthy salads. There are actually two different species and they look distinctively different. Summer purslane (*Portulaca oleracea*) is a succulent with short, thick stems and sturdy leaves. It is exceptionally nutritious, as it is an excellent source of omega-3 fatty acids (in particular alpha-linolenic acid) and contains more eicosapentaenoic acid (EPA) than any other terrestrial plant. Moreover, it is rich in calcium and contains potassium, magnesium, and iron, as well as an abundance of antioxidants. Winter purslane (*Claytonia perfoliata*), also called miner's lettuce, is more delicate with a long, thin stem and a small, round leaf. It is very rich in vitamin C. At the end of the season toward spring, the stems of winter purslane are very long and a tiny white flower forms on top of the leaves. Both the stems and the flowers are edible and should not be discarded.

Purslane salad with seaweeds

2g dried giant kelp or bullwhip kelp

500ml fresh purslane

1 blood orange

Separate the larger purslane plants into smaller pieces and place them in a bowl. Peel the blood orange and cut it into small cubes that are mixed gently into the purslane. Toast the seaweed blades and crush them. Sprinkle over the salad and serve immediately while the purslane is still crisp and before the orange juice has softened it.

MARINATED WINGED KELP

This recipe was given to me by Rae Hopkins, who has the good fortune to live on the Pacific Coast of Vancouver Island, where she herself can harvest fresh winged kelp (*Alaria marginata*). Here, however, the recipe calls for it in dried form. If a milder taste is desired, *arame* can be substituted for winged kelp.

Marinated winged kelp   *serves 4 as an appetizer or a salad*

250ml dried winged kelp cut into strips

125ml rice vinegar

3 tsp soy sauce

1 tsp sugar

1 tsp salt

Soak the seaweed strips in cold water for about 15 minutes. Drain the water and cut away the thick, tough mid-rib. Place the seaweeds in a bowl together with the other ingredients and allow them to marinate for 10 minutes. Serve as an appetizer on wholemeal crackers or place the strips on top of a green salad together with a few pieces of thinly sliced spring onion.

CUCUMBER SALAD WITH SEAWEEDS

This salad is a Japanese-inspired version of the traditional Danish cucumber salad, which derives its taste from a sweet and sour dressing. It is easy to make and keeps for a couple of days in the refrigerator.

The blades of soaked, dried *wakame* have a tendency to stick to each other when wet. For this reason, they should be carefully mixed with the cucumber slices.

Cucumber salad with *wakame*   *serves 4*

4g dried *wakame*

1 cucumber

salt, sugar

rice vinegar

soy sauce

Cut a cucumber into thin slices, place it in a bowl, and sprinkle with salt. After about 10–20 minutes the salt will have drawn a great deal of liquid out of the cucumber slices. Remove them from the bowl and squeeze them in your hands. In the meanwhile, the seaweeds have been placed in water to soak and then sliced into squares that are about 3cm across. Prepare a dressing from rice vinegar, soy sauce, sugar, and salt and mix with the seaweed pieces and the cucumber slices. Add salt and sugar to fine-tune the taste, if required.

WAKAME* SALAD WITH *TOFU

Somewhat firm *tofu* is a good addition to a seaweed salad. The white of the *tofu* contrasts well with the color of the seaweeds to create an appealing dish.

Wakame salad with *tofu*   *serves 4*

5g dried *wakame* cut into 2–3cm squares

250g *tofu*, diced

250g coarsely grated carrot

1 cucumber, diced

For the dressing:

2 tbsp rice vinegar

2 tbsp soy sauce

2 tbsp marinade from pickled ginger

Prepare the dressing. Meanwhile, soak the dried seaweeds in cold water for 10 minutes and then drain the water. Mix the *wakame* with the *tofu*, vegetables, and dressing. Add salt to taste and sprinkle toasted sesame seeds on top.

SIMMERED *HIJIKI*

Hijiki is a versatile brown alga, but, as it tastes somewhat bland on its own, it is best to combine it with other ingredients. An ideal way to add taste substances is to simmer it in the soup stock *dashi*, which will impart a slightly smoky flavor to the seaweed. This salad will keep in the refrigerator for a couple of days.

▲ Dried *hijiki*.

Simmered *hijiki* and carrot salad   *serves 4*

8g dried *hijiki* or *arame*

1 carrot, julienned

75ml *mirin* (sweet rice wine)

50ml *dashi* (Japanese soup stock)

soy sauce, sugar

▸ A dried specimen of the red alga *Claudea elegans* from the collections of the Natural History Museum in London.

Soak the dried *hijiki* in cold water for about 30 minutes. Drain the water and wash the seaweed thoroughly several times in clean water. Mix the seaweed with the carrot strips and add the *mirin* and *dashi*. Simmer over a very low heat until most of the liquid is gone. Adjust the taste with soy sauce and sugar. Allow to cool before serving.

Insula Van Diemen, Georgetown — Dr Harvey. No. 145 S.

QUINOA SALAD WITH DULSE

Quinoa, which is neither a true cereal nor a grass, is actually closely related to beets and spinach. Seeds from this plant were a staple in the diet of the ancient Incas of South America in the same way as rice was a basic foodstuff in Asia. Quinoa seeds contain significant amounts of proteins and fats, especially the unsaturated omega-3 fatty acids such as alpha-linolenic acid—they are considered exceptionally nutritious. The recipe below is for a little side dish that combines cooked quinoa with dulse, which contributes both color and flavor.

Quinoa salad with dulse   *serves 6*

10g dried dulse, in small flakes

250g quinoa

600ml water

soy sauce

Soak the dulse flakes in water to which a little soy sauce has been added. Rinse the quinoa thoroughly in cold water, then roast them at low heat in a pot without any fat, until they turn golden. This helps to remove some of the bitter substances found in the seed coverings. Add the water and cook over low heat for about 20–25 minutes, being careful not allow the pot to boil dry. Drain the seeds and let them cool. Remove excess liquid from the rehydrated dulse flakes and mix them into the quinoa.

▲ Dried dulse (*Palmaria palmata*).

SALADS AND SAUCES WITH SEA SPAGHETTI

Thongweed (*Himanthalia elongata*) is a brown alga, found along some Atlantic coasts from Portugal to the Faroe Islands. The seaweed has a cup-shaped vegetative base that looks like a button with a short stipe. From the base grow long, forked receptacles formed as ribbon-like blades that are up to 1.5cm wide. These receptacles, which carry the reproductive structures, resemble leather thongs and can be up to several meters in length. Because of its characteristic appearance, thongweed is also called sea spaghetti. In France it goes by the name of *haricots de mer* (green beans from the sea) because the individual pieces also look somewhat like longbeans. Thongweed can be bought in packages as dried strips or in jars as a brine-pickled preserve.

▲ Thongweed (*Himanthalia elongata*).

Salad with sea spaghetti   *serves 4*

10g dried or 40g preserved thongweed

1 cucumber

2 tomatoes

1 large avocado

1 small bunch radishes

juice of 1 lemon

fresh basil, chopped finely

soy sauce and olive oil

1 tsp mustard

Seaweeds in salads and sauces

If using dried thongweed, soak it in water for half an hour; if using preserved thongweed, simply drain off the liquid. Slice the cucumber, place the slices in a bowl with a little salt for about 10 minutes and then squeeze out any excess moisture. Cut the avocado and tomatoes into small pieces and the radishes into thin slices. Combine these ingredients and make a dressing with mustard and the juice of the lemon, adding soy sauce and olive oil to obtain the desired taste and consistency. Toss the salad and gently add in the chopped basil.

Thongweed can also be eaten like a pasta with a tomato sauce or mixed into a tomato marinade, which is superb on boiled potatoes.

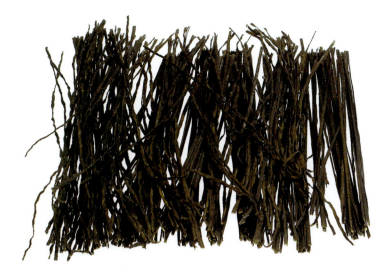

◄ Dried thongweed (also called sea spaghetti).

Sea spaghetti with tomato sauce   *serves 4*

30g dried thongweed or about 120g preserved thongweed

200ml sun-dried tomatoes

3 sun-ripened tomatoes

200g button mushrooms

1 clove garlic, chopped

6 black olives, finely chopped

avocado

pine nuts

Soak the dried thongweed in water for about half an hour, then cook it for 10 minutes. Drain and rinse thoroughly in cold water. If using preserved thongweed, simply drain it. Mix the tomatoes, garlic, mushrooms, and olives together in a pot and allow the sauce to cook gently for about 5–10 minutes. Serve the cooked seaweed with the sauce and garnish with avocado and pine nuts.

SEAWEEDS IN DRESSINGS

Seaweeds can add zest, taste, and color to an otherwise dull salad dressing.

Yogurt dressing with seaweeds

1 tbsp dried flakes of *nori*, dulse, winged kelp, or *wakame*

200ml plain yogurt, drained

1 tsp sugar

Whisk together the ingredients and serve on a green salad or with a fish dish.

▸ Preserved thongweed.

GUACAMOLE WITH DRIED SEAWEEDS

Guacamole is a Mexican dish consisting of puréed avocado to which lemon or lime juice has been added. It can be used as a dip or as an appetizer.

Seaweeds in salads and sauces

Guacamole with dried seaweeds   *serves 4*

5 tbsp dried sea lettuce or dulse, chopped finely or in flakes

4 ripe avocados

juice of 1 large lemon

3 sun-ripened tomatoes

1 onion

2 cloves of garlic

parsley, chopped

olive oil

Peel the avocados and remove the pits. Mash the avocados with a fork and add the lemon juice. Chop the tomatoes and onion coarsely and put the garlic cloves through a press. Stir everything together, adding the seaweed and chopped parsley. Adjust the seasoning with a little powdered white pepper and olive oil. This can be served on toasted bread.

AVOCADO SALAD WITH BLADDER WRACK

Bladder wrack provides texture and color to a summer salad. Blanched top shoots can be mixed into any green salad, e.g., with avocado.

Avocado salad with blanched bladder wrack   *serves 4*

20 top shoots of freshly harvested bladder wrack

3 avocados

juice of one lemon

6 shredded marinated ginger slices in 3 tbsp marinade

¼ tsp *wasabi* powder

Blanch the bladder wrack shoots for a few seconds in boiling water and transfer them immediately to ice water. Drain the cooled bladder wrack and let it marinate in the lemon juice for 10 minutes. Peel the avocados and cut them into cubes. Chop the ginger finely and mix it together with the ginger marinade and the *wasabi*. Gently coat the avocado and bladder wrack with the mixture.

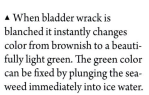

▲ When bladder wrack is blanched it instantly changes color from brownish to a beautifully light green. The green color can be fixed by plunging the seaweed immediately into ice water.

Seaweeds in omelettes and in fish and vegetable dishes

OMELETTE WITH THREE TYPES OF SEAWEEDS

It is easy to add different types of seaweeds to one's favorite omelette recipes. The algae contribute a salty taste and provide a beautiful color contrast to the yellow of the eggs.

Omelette with three types of seaweeds	*serves 4*

5g dried sea lettuce
5g dried dulse
3g dried *Porphyra* or a sheet of *nori*
4 eggs
parsley

Soak the seaweeds. If *nori* is being used instead of *Porphyra*, chop the sheet into smaller pieces. Whisk the eggs together, possibly with a little cracked pepper and a bit of parsley. Chop the rehydrated seaweeds into smaller pieces and stir into the egg mixture. Cook the omelette in the conventional way on a large frying pan.

OVEN-BAKED SALMON WITH DULSE

Dried, toasted dulse adds a nutty taste, with nuances of fresh seawater, which goes well with just about any type of fish dish. This recipe is easy and quick to follow; using dulse that has been smoked over applewood gives it a special twist.

Oven-baked salmon with dulse	*serve 4*

5g dried, toasted dulse, preferably applewood smoked
2 thick salmon steaks, cut in half lengthwise and boned

Place the salmon fillets in an ovenproof dish or on baking paper in a roasting pan. The fillets can be with or without skin, but they will have more flavor if the skin is left on and they are cooked skin-side down. Bake in a 200°C oven for 10–15 minutes, depending on their size, thickness, and whether the fish is to be cooked through. When the pieces are done, remove them from the oven and sprinkle the

◄ Oven-baked salmon with dulse, served here with green asparagus and *hijiki*.

toasted dulse over the fillets. Serve with rice, vegetables, oven-dried aubergines, or an avocado salad. One can also serve the salmon with simmered *hijiki* (see recipe above).

Tuna and seaweed salad

Canned, water-packed tuna is not particularly tasty on its own, but seaweed comes to the rescue by adding salt and *umami* taste.

Tuna and seaweed salad	*serves 4*

5–10g dried, toasted *wakame*, *nori*, dulse, giant kelp, or bullwhip kelp
2 cans water-packed tuna (ca. 300g)
1 small spring onion, finely chopped
1 green apple, diced
lemon juice
mayonnaise

Drain the water from the tuna; break it up into small bits in a bowl. Crush the seaweed in your hand and add, together with the spring onion and apple, to the tuna. Stir the mixture together with salt, pepper, and lemon juice. Add mayonnaise according to taste and desired consistency. The salad can be presented on its own as a small luncheon dish, served on French bread, or used as filling in pita pockets.

Carpaccio with bladder wrack

This recipe was given to me by Klavs Styrbæk. He marinates the youngest and most delicate shoots of ordinary bladder wrack, which is far too hard and tough to eat as is.

Carpaccio with bladder wrack

fresh shoots of bladder wrack (blades and bladders)
fresh cod or turbot fillet
canola oil
lemon/lime
sea-buckthorn berries, crushed
cayenne pepper

Blanch the seaweed, bringing out its pale green color. Place it on the fish with the grated rind of a lime, cayenne pepper, and crushed sea-buckthorn berries. Roll up the raw fish and freeze it. To serve, slice it thinly and marinate the pieces in salt, canola oil, and citrus juice. The seaweed shows up as green patterning against the white of the fish.

Seafood salad—now with seaweeds, too

A salad made with fish and shellfish is a delectable dish, which can be made even better by adding seaweeds to it. The recipe below is for an easy to prepare, infinitely variable salad that incorporates a whole range of 'treasures from the deep'.

Shellfish and seaweed salad   *serves 4*

5g dried *wakame*, giant kelp, or bullwhip kelp
150g cooked, peeled shrimp
1 small cucumber, diced
150g seared scallops, marinated calamari, cooked razor shell clams,
    cooked cockles, or any combination of these
rice vinegar and soy sauce

Soak the seaweeds in water and drain. Mix together the ingredients in a bowl and drizzle a rice vinegar and soy sauce dressing over them.

CAVI-ART

Cavi-art is the catchy name given to a series of products invented by a Danish firm, Jens Møller Products, that turns seaweeds into small, round particles, which, as the name implies, resemble fish roe. The 'fish eggs' come in different colors and sizes, corresponding, for example, to red trout roe and black lumpfish roe. The basic ingredients in Cavi-art are granules from a blend of three different brown algae, which are then reconstituted in an alginate gel. How this actually works is a trade secret, but the company states that the original nutritional components of the seaweeds are retained in the finished product.

Nevertheless, Cavi-art does not need to pretend to be something else. The different varieties are excellent in their own right as an innovative seaweed product and can be enjoyed in that way. Cavi-art is distributed widely in Europe and has made its way into the kitchens of professional North American chefs.

Seaweeds in omelettes and in fish and vegetable dishes

▲ Black and red Cavi-art.

Fish mousse with Cavi-art   *serves 4*

100ml red or black Cavi-art

550g cooked trout or salmon fillet

1 large onion

500ml crème fraiche

12 gelatine leaves (or 2 sachets, about 10g of powdered gelatine)

100ml thick cream, whipped

juice of half a lemon

salt and pepper

Blend together the fish, crème fraiche, onion, and lemon juice. Stir in the Cavi-art, together with salt and pepper. If it is not possible to obtain Cavi-art, one can use *konbu* by first rehydrating it thoroughly and then blending it. Mix in the gelatine, which has been dissolved according to directions on the package. Fold in the whipped cream, pour the mousse into a ring mold, and refrigerate.

MASHED POTATO WITH DULSE

Dulse and potatoes are a perfect match for each other. It is quite likely that in earlier times, especially in coastal areas, dulse was used to pep up an otherwise uninteresting potato dish during the long winter months when fresh vegetables and herbs were both in short supply.

Mashed potatoes with dulse *serves 4*

8g dried dulse
1kg potatoes
butter

Soak the dulse in water for about 10 minutes, drain it, and chop it into small pieces. Peel, cook, and mash the potatoes with a little butter. Stir the dulse into the warm potato mash and add salt and pepper to taste. This dish goes well with smoked meat.

▸ Dried dulse from Mendocino, California.

LAVERBREAD—A TRADITIONAL WELSH DISH

Purple laver (*Porphyra umbilicalis*) is found in many places along the coasts of Wales, Scotland, and England. In accordance with the traditional recipe, fresh purple laver is boiled for several hours in lightly salted water, until the blades of the seaweed break up into small pieces. When the excess liquid is drained away, what remains is a thick, spinach-like purée, which is referred to as laverbread. Warm laverbread can be spread on toast and, if desired, garnished with crisp, fried bacon.

To make a classical Welsh breakfast, heat the laverbread in the juice and fat from fried bacon. Eat it with bacon, mushrooms, and eggs. Another traditional breakfast recipe mixes laverbread with oatmeal to form small flat cakes that are fried and served with crisp bacon and cockles. The dish can be seasoned with lemon juice and pepper.

GREEN BEANS WITH KELP

Green beans and seaweeds can be said to be the best, most basic, and healthy combination of foods that grow on land and in the sea. The beautiful colors of the dark kelp and the green beans enhance a dish that is easy to prepare and that can be eaten both hot and cold. Experience shows that the beans need a shorter cooking time when they are boiled together with the kelp.

Seaweeds in omelettes and in fish and vegetable dishes

◄ Dried sea tangle (*Laminaria digitata*).

Green beans with kelp   *serves 4*

10g dried kelp (*konbu*, winged kelp, or sea tangle) cut into 2–3cm
    long pieces
500g fresh green beans

Soak the seaweeds in water. If no *konbu* is used, this step can be omitted. Bring water to a boil and cook the beans and the seaweeds for ca. 5 minutes. Serve the dish as a complement to meat and fish dishes.

GREEN ASPARAGUS WITH SEAWEEDS

Green asparagus, like green beans, can be prepared or served with a brown alga such as *wakame*, bullwhip kelp, or *hijiki*, which will add flavor, salt, and color to the dish.

Welsh caviar—laverbread

Laverbread, made from the European species *Porphyra umbilicalis*, popularly known as purple laver, can best be described as a spinach-like, black purée of cooked, chopped seaweeds. In Europe, this way of using seaweeds is probably the one most steeped in tradition and it can be considered the closest analogy in the West to Japanese *nori tsukudani*.

Traditionally, laverbread (*bara lawr*) is eaten for breakfast, heated in bacon drippings, and often served with cooked cockles. In earlier times, oatmeal was added to the seaweed purée to make the mixture firmer and to make it go a little further. Older people in Wales regard laverbread as a delicacy and, without the slightest trace of irony, refer to it as Welsh caviar. Nevertheless, it is likely that, throughout the ages, laverbread has been regarded as a sort of subsistence food that one could always fall back on when times were hard.

In Wales the laverbread tradition is still honored, but it is produced in only a few places and the volume of sales is declining, probably because it is now eaten mostly by the elderly. Fresh laverbread can be bought in the southwestern part of Wales; elsewhere it is mostly available only in cans. If one has grown up eating laverbread for breakfast, however, it is difficult to do without it and, therefore, the famous London department store, Harrods, carries the fresh product, proudly supplied by Selwyn's Penclawdd Seafoods.

Selwyn's Penclawdd Seafoods is a small, family-run business located in the village of Llanmorlais, near Penclawdd on the northwest coast of the Gower Peninsula facing the broad Loughor Estuary. Until the beginning of the 1900's, Penclawdd was a busy port, whose thriving economy was based on the mining industry and tinplate, copper, and brass works. For centuries, right back to Roman times, the town has also been associated with cockle gathering. For the widows of the miners who had, as the saying went, died of 'dust in the lungs', this activity was truly a lifesaver. And after the coal mine closures in the 1980's, the cockle harvest was one of the few remaining ways of earning a living in Penclawdd.

This was also the case for Selwyn's mother, Sarah Jones, who was widowed at an early age and taught her seven-year old son to gather cockles.

They brought them to the market at Swansea and sold them, as well as laverbread. Seaweeds were collected on the south coast of the Gower Peninsula, where the sand and stone beaches of the Bristol Channel created ideal growing conditions for *Porphyra umbilicalis*. Cockles were dug from the sandbanks of the river estuary. So in a sense, the particular combination of laverbread and cockles is an interesting result of the close proximity of these two geographically distinct areas located within a small part of southwest Wales.

I had the opportunity to visit Selwyn's on a trip to Wales. In the morning I met Selwyn's son Brian, his wife Alyson, and their son Ashley, who are now trying to expand their business. Ashley has gone to Japan to promote their laverbread and to investigate whether it might be feasible to market it to the rather discriminating Japanese consumers. Seemingly, the Welsh *nori* was well received. Ashley's impression is that, although the Japanese feel that Selwyn's laverbread is of good quality, it perhaps tastes a little bit different from Japanese *nori*. Besides, the product will not, under any circumstances, be marketable in cans as the Japanese will consider it only if it is presented in glass jars or in plastic packages. A Japanese *nori* expert subsequently visited the company in Wales to study the quality of the raw materials and the ways in which they are processed. Ashley now has his doubts about whether Selwyn's can make any inroads in Japan. In the meanwhile, he has decided not only to sell his laverbread in cans, but also to use clear plastic packaging that will allow the customer to see the contents.

Brian Jones told me that during the seaweed season Selwyn's produces one ton of laverbread a week. For each ton of finished laverbread, they use a half ton of fresh algae. Previously, several families were engaged in harvesting the algae, but sales are on the downswing and now only a single family is involved. Brian thinks, however, that it is possible to harvest much greater quantities in a sustainable manner. The seaweeds grow quickly and in the summer one can harvest in the same area after a two-week interval. The harvest takes place at low tide, with the gatherers walking directly on the seabed and cutting the seaweed plants or pulling them up by hand. They are collected, bundled in nets, and sent to the factory still wet. Here they can be kept for up to a week in a cool room before being processed.

▲ Harvesting purple laver (*Porphyra umbilicalis*) at low tide on the coast of Wales.

▲ An old photograph of Welsh 'cockle women' cleaning the cockles that are eaten with laverbread.

The actual processing is very straightforward. First the seaweeds are washed in water to remove sand, stones, and bits of shell. The algae are then cooked for 4–5 hours, after which they are cooled and chopped into small pieces. A little salt is added to the resulting purée. Some of the fresh laverbread is packaged in small plastic containers for sale in the immediate area as a fresh product that will keep for about a week. The rest is canned for wider distribution and has a shelf life of about a year.

None of this is very different from the time-honored ways. The miners' wives gathered and cooked the fresh seaweeds, wrapped them in a cloth to allow the excess liquid to drain off, and carried the finished laverbread to market in a basket or shared it out within their local circle of mine worker families. Interest in laverbread and its distinctive taste spread from these small mining communities to other parts of Wales. Brian said that these families were convinced that laverbread was not only a subsistence food, but that it also had something to do with well-being. It was common knowledge that it helped those who suffered from goitre and that the health of sick miners improved when they ate laverbread. It is Brian's opinion that there was a folkloric wisdom from ancient times that was put into practice to deal with the diseases that came in the wake of the Industrial Revolution. The miners drank the salty liquid left over from cooking cockles and seaweeds and used it to rinse away the coal dust that settled in their throat and pharynx.

Five generations of the same family have now been making laverbread at Selwyn's and even more generations of other families before them in Penclawdd. But what the future holds is uncertain. As Brian Jones said, "Young people are not eating laverbread."

"And where might one buy fresh laverbread?" I asked. "From the butcher or the fishmonger," Brian replied. "From the butcher you get the bacon in which the laverbread is fried and from the fishmonger the cockles that are the traditional Welsh accompaniment for laverbread. Otherwise, you can buy it from us, which is cheaper!" I left Selwyn's with a couple of containers of laverbread, a European seaweed food with a rich tradition.

▲ Laverbread.

▶ ▶ Scenes from the Penclawdd area in southwest Wales where a small family-run business, Selwyn's Penclawdd Seafoods, produces laverbread.

Green asparagus with seaweeds   *serves 4*

5g dried *wakame*, giant kelp, or bullwhip kelp
1 bunch fresh green asparagus, not too thick
100ml water
1 tsp butter

Trim the ends of the asparagus to remove the tough parts. Then
cut 2–3mm thick slices from the bottom 1–2cm of the stalks. Bring
the water to a boil in a pan and add the butter. Place the stalks and
the slices in the water and sprinkle the seaweeds, which have been
crushed into pieces that are not too small, on top. Simmer for about
10 minutes, until most of the water has evaporated. Serve as an accompaniment to a fish dish.

Green asparagus with *hijiki*   *serves 4*

8g *hijiki*, simmered as described for simmered *hijiki* salad above,
    but without the carrot
1 bunch fresh green asparagus

Trim the asparagus and steam for about 5 minutes. They should still
be slightly crisp. Arrange them on a platter and spread the simmered
hijiki on top. They can be served with a dish of steamed fish.

STEAMED ZUCCHINI WITH SEAWEEDS

Lightly roasted or steamed zucchini often have an uninteresting, bland taste,
which can be remedied with the addition of a little seaweed. *Nori, wakame,* or
dulse can work wonders. Steamed zucchini, especially the small pale green
varieties, go well with a fish dish or can be served cold as a salad.

Steamed zucchini with seaweeds   *serves 4*

toasted *wakame*, bullwhip kelp, *ao-nori,* or *furikake* with *nori*
6 small zucchini
30g pickled ginger, finely chopped

Wash the zucchini, trim off the ends, and cut into slices at a 45° angle,
turning the zucchini 90° after each cut. Large zucchini can be cut into
long thin strips. Place the zucchini and the ginger in a pot or pan with

▸▸ Dried bullwhip kelp,
shown here with a little
ao-nori sprinkled on top.

154

a very small amount of boiling water, as well as a little of the juice from the pickled ginger, if desired. Steam the zucchini for about 2–3 minutes, turning them carefully a couple of times. Do not let them brown. If necessary, place a lid on the pot for some of the time—the longer it is on, the more the zucchini will steam through on account of the liquid they give off. Do not let them get too soft. Place the steamed zucchini in a bowl and sprinkle crushed, toasted seaweeds or *furikake* over them. Pay attention to the delightful aroma given off by the dried seaweed flakes as they absorb moisture and warmth from the zucchini. Serve the dish immediately before the wonderful smell has dissipated.

GREEN LENTILS WITH *WAKAME*

Lentils, for example, the small ones from Puy, contain large amounts of proteins and vitamin B and are especially rich in polyunsaturated fats. When combined with seaweeds, lentils make a healthy, tasty vegetarian dish, which brings together some of the best elements from the land with some of the best from the sea.

Green lentils with *wakame*	serves 4

15g dried *wakame* or winged kelp
250g green lentils, rinsed very carefully and checked for tiny stones
50g shallots
50g dark raisins, plumped
3 laurel leaves
herbs
olive oil
soy sauce

Soak the seaweeds for 20 minutes and then cut them into smaller pieces. Remove the mid rib, as required. Sear the shallots and the herbs in a little olive oil for a short time, then add the rest of the ingredients, other than the soy sauce. Allow to simmer for about 15–20 minutes and adjust the seasonings with soy sauce.

Seaweeds and sushi

Seaweeds and sushi

The red alga *Porphyra*, in dried and toasted form, is the only type of seaweed used for making sushi. All sushi is based on cooked rice, flavored with rice vinegar, salt, and sugar. This sour-sweet rice is combined with fish, shellfish, vegetables, omelette, or roe. *Nori* is used in different ways in sushi to add taste, color, and texture. Sheets of *nori* are needed to give support and structure to *maki*-zushi and *gunkan*-zushi and small strips sometimes tie pieces of fish, omelette, and shellfish to the hand-formed rice balls in *nigiri*-zushi.

ONIGIRI—SEAWEEDS ON THE GO

The very simplest Japanese brown-bag lunch consists of *onigiri*. This is plain cooked rice formed into a decorative triangular or oval shape, which is wrapped in a sheet of *nori* and usually dipped in soy sauce as it is being eaten. There is often some filling inside the rice ball, for example, pickled *shiso*, *umeboshi*, *nattō*, cooked tuna, or a little seaweed salad.

Onigiri can be bought ready made and wrapped in plastic in Japanese kiosks and convenience stores. The sheet of *nori* is sealed off in its own package to keep it dry and preserve its crispness. As the seaweed quickly absorbs moisture from the rice ball and becomes soft, it is wrapped around the rice just before it is to be eaten.

Japanese vocabulary: *Shiso* is a pungent green herb (*Perilla frutescens*). *Wasabi* is Japanese horseradish (*Wasabi japonica*). *Nattō* is made from fermented soy beans. *Umeboshi* are salt-pickled Japanese plums (*Prunus mume*). 'Sushi' and 'zushi' have the same meaning, but are pronounced with an s and a z sound, respectively. In Japanese, sushi is pronounced with a voiced z when linked to another word. For this reason, '*nigiri*-sushi', for example, is pronounced and written as '*nigiri*-zushi.'

▲ *Onigiri*—sushi rice wrapped in sheets of *nori*.

Rolled sushi with *nori*: *maki-zushi*

The most widespread use of *nori* for sushi is *maki*-zushi, which are sushi rolls made with a whole or a half sheet of *nori* to hold them together. Without a doubt, this is by far the way that the majority of people in Western countries have been introduced to seaweeds as something edible and tasty.

Maki-zushi is made from a toasted sheet of *nori* (*yaki-nori*) rolled around sushi rice and a variety of fillings with the help of a rolling mat made from thin strips of bamboo. The rolls can be thin (*hosomaki*) or thick (*futomaki*) and the *nori* can be on the outside or on the inside with the rice on the outside (*uramaki*). They can be made with just about any type of filling—fish, shellfish, mushrooms, omelette, and vegetables, either on their own or in a combination of different textures and colors. These can be arranged in such a way that beautiful patterns, symbols, or pictures are created that show up when the rolls are sliced. The variety is limited only by the imagination of the chef.

Yaki-nori used for making *maki*-zushi is toasted and wrapped in airtight packages to keep it dry and crisp. Because it absorbs liquid very easily and becomes soft and fragile, it is important to work quickly. When making *maki* rolls, it is also necessary to find a balance between avoiding touching the *nori* sheets with damp fingers and keeping the fingers moist so that rice and filling do not stick to them when the roll is being assembled.

For thin *maki* rolls (*hosomaki*) a half sheet of *nori* is used and for the thick ones (*futomaki*) a whole one is needed. *Hosomaki* are typically 2.5 cm in diameter and *futomaki* about 4–5 cm.

Maki-zushi

1 or 1/2 sheet of *nori*, as per above
cooked sushi rice
filling as described above
wasabi

Place a sheet of *nori* on the flat side of the bamboo rolling mat with the shiny side down and one end of the sheet lined up with the end of the rolling mat that is closest to you. Put a suitable amount of cooked sushi rice in a wide strip on the sheet and then distribute it evenly all over the sheet, leaving uncovered only about 2–3 cm at the end that is away from you. First, using your finger tips, spread a little

Sushi ingredients: For cooking sushi rice and for advice regarding the selection of raw and prepared fish and shellfish for sushi, consult an appropriate cookbook, e.g., the author's *Sushi. Food for the eye, the body & the soul* (Springer, New York).

wasabi on the rice at the end nearest to you. Next, place the filling lengthwise on top of the *wasabi*. Now, holding the mat with both hands use it to roll the whole works together, exerting a light, even pressure as you roll forward. At the same time, as the mat is progressively uncovered hold it up and away from the sushi. As you get to the far end, there should be a little overlap where there is only the *nori* sheet. After a few moments, the moisture in the rice will cause the two *nori* surfaces to stick together and the end of the sheet will lie tightly against the sushi roll, forming a seal. If any filling or grains of rice stick out at the ends of the roll, you can push them inside or trim them away. In some cases, though, the bits that are poking out can have an extra, decorative effect. Once the roll is finished, it is cut up before being served. The knife must be spotless, very sharp, and moistened in order to produce a clean cut. Thin rolls are normally sliced into six equal pieces, about 3–4 cm long, which are served in a group arrangement. Thick rolls are usually cut into thinner slices that are about 1.5 cm thick.

A particularly sophisticated form of *futomaki* rolls consists of rolls within rolls. This is a way to create patterns and pictures, using *nori* and filling, that appear when the rolls are sliced crosswise. For a simple version, press the thin roll into a triangular or square shape, or possibly that of a tear-drop with one sharp edge. Place the *hosomaki* on the rice of the thick roll and add more filling as desired.

Inside-out rolls (*uramaki*), which can be both thick and thin, are made more or less in the same way as the ordinary *maki*-zushi described above. The most important difference is that the rolling mat has to be covered in plastic wrap that is carefully folded around the edges of the mat so that the film clings to itself.

◄ *Maki*-zushi—*hoso-maki* and *futomaki*.

Uramaki

Place the sheet of *nori* on the mat as if making ordinary *maki*-zushi and spread the rice uniformly over the whole surface. Then, with a quick motion, take the sheet and flip it over so that the rice is against the surface of the mat. The plastic film will prevent the rice from sticking to the mat. Place *wasabi* and filling on the *nori* and use the mat to roll the *uramaki* together. As the rice is now on the outside, it is not difficult to make the seam of the roll seal properly. Before the roll is sliced, it can be sprinkled with, or rolled in, sesame seeds or fish roe.

A particular type of large *uramaki* roll has a decorative pattern made from individual thin pieces of fish of different colors, avocado, or perhaps kiwi, which are placed on and pressed into the surface of the roll after it has been made. Some of these rolls are known as rainbow rolls (*tazuna*-zushi) because the interplay of colors, with stripes of red, green, and white, are reminiscent of a rainbow.

Maki-zushi can also be made using rehydrated *wakame*, giant kelp, or bullwhip kelp instead of *nori*. It is important that the blades are thin and not too tough. Bullwhip kelp has the advantage that it secretes only a little bit of slimy substance when it is soaked in water.

In rare cases, *nori* is also used to make *sashimi*, which is completely raw fish. A sheet of *nori* is wound around a strip, either square or roundish, of deboned fillet of fish, for example, red tuna, salmon, or a white fish. The roll is then sliced crosswise in pieces that are about 1 cm thick and eaten like *sashimi* with soy sauce and *wasabi*. The black *nori* surface is a fine contrast with the red or the white of the fish.

▶ *Sashimi* (raw fish) with tuna wrapped in *nori*.

Hand-rolled sushi with *nori: temaki-zushi*

Hand-rolled sushi is called *temaki*-zushi and comes in several shapes, the most common of which is a cone. The cone is made from half a sheet of nori and can be filled with different types of fish, shellfish, roe, grilled fish skin, crabmeat, avocado, cucumber, omelette, green salad leaves, *shiso* leaves, and so on. *Temaki*-zushi can also be rolled into a cylindrical shape. To do this, place a small piece of *nori* going crosswise to the bottom of the large sheet of *nori* to form a sort of U-shaped base. This prevents the filling from falling out at the bottom.

Seaweeds and sushi

◄ *Temaki*-zushi.

Sushi cone (*temaki*-zushi)

1/2 sheet of *nori*
cooked sushi rice
filling as described above
wasabi

Place the half sheet of *nori* in the left hand (if you are right-handed) and put a clump of cooked sushi rice in the middle of the sheet. Spread a little *wasabi* on the rice with your fingertips. Place as much filling as the cone will be able to hold on the rice. The cone is now closed by folding the left side in over the rice and rolling it into a cone. To finish it off, you can decorate the *temaki* by putting a spoonful of roe on the filling.

Temaki-zushi is a finger food and should be eaten immediately after it has been prepared before the *nori* has a chance to absorb moisture from the rice and while it is still crisp and delicious. Hand rolls that have been lying around for too long are often chewy because the sheet of *nori* has become too soft.

BATTLESHIP SUSHI WITH *NORI*: *GUNKAN-ZUSHI*

Battleship sushi, or *gunkan*-zushi, is a type of *maki*-zushi made with filling that can be difficult to use on *nigiri*-zushi or sushi rolls because it does not stick together well or is too soft or too moist. A little strip of *nori* is fastened around a clump of rice, extending above the top of the rice, so that the resulting 'boat' can be loaded with filling.

Gunkan-zushi

1 sheet of *nori* cut into strips ca. 4 cm×15 cm
cooked sushi rice
filling, for example, roe, oysters, scallops, marinated *shiitake*
wasabi

Shape a little clump of cooked sushi rice into a round or elongated ball and carefully fold the *nori* strip around it. The seaweed sheet should rise about 1 cm above the top surface of the rice. To fasten the *nori* strip together, place a single grain of the sushi rice between the end of the strip and the shaped sushi. Place the piece on a flat surface and press the rice together gently so that it looks like a little boat (hence, battleship) that can carry a cargo of filling that goes up to the edge of the *nori*, or even a bit higher. Spread a thin layer of *wasabi* on top of the rice and add the filling. Possibilities include finely chopped pieces of fish, roe, oysters, scallops, and so on. The finished *gunkan*-zushi can be decorated with a little bit of greenery, for example, cucumber, avocado, green *shiso*, or mild garden cress.

► *Gunkan*-zushi.

162

CHIRASHI-ZUSHI WITH STRIPS OF NORI

Chirashi-zushi is a really easy way of preparing sushi. In Japanese *chirashi* means to spread. The simplest way to make it is to place a variety of toppings on a layer of cooked sushi rice in the bottom of a flat-bottomed bowl. Typically *chirashi*-zushi should be in the proportions of two-thirds rice and one-third toppings. Seaweeds enters into the picture as a thin layer of finely cut *yaki-nori* strips are sprinkled over the rice before the topping is added.

Chirashi-zushi   *serves 1*

1 sheet of *nori* cut in fine strips, ca. 0.2 cm × 3 cm
cooked sushi rice
roe, for example, lumpfish roe
toppings, for example, fish, shellfish, omelette, mushrooms, vegetables
wasabi

Place a layer of sushi rice in a flat-bottomed bowl with low sides. Press the rice gently together with the top of the outer part of the fingers of one hand held together. The layer of rice should be firm, but not too firm. Make sure the fingers are moist so that the rice does not stick to them. Spread a thin layer of roe on top of the rice. Using bone-dry chopsticks, add a layer of *nori* strips and then work quickly to place the chosen toppings in an artistic arrangement. Alternatively, the *nori* strips can be sprinkled over the *chirashi* just before serving.

OMELETTE WITH NORI

Omelette (*tamago-yaki*) is used as a filling or topping for all types of sushi. The omelette is made by folding the lightly cooked egg over and over into a little multi-layered package that can be cut into pieces or strips, as required. By placing a sheet of *nori* on top before the folding process starts, it is possible to create a beautiful pattern of black filaments in the yellow omelette. This shows up when the omelette is cut crosswise.

Omelette (*tamago-yaki*) with *nori*

1 sheet of *nori*
3 eggs
mirin (sweet rice wine)
salt and sugar

▲ *Nigiri*-zushi with omelette (*tamago-yaki*) fastened with a ribbon of *nori*.

Crack the eggs into a bowl. Add a little salt, sugar, and *mirin* (optional) and whisk everything together lightly with a fork. Heat a pan that has been greased with a tiny amount of fat, preferably one that has virtually no flavor of its own. Pour the egg mixture into the pan a little at a time over low heat. Place the *nori* sheet on the surface and, using chopsticks or a wooden spatula, fold the set egg mixture together on itself several times to create a flat, layered omelette. Because the sheet of *nori* will shrink very rapidly, the folding process has to be carried out quickly so that the seaweeds do not end up as a lump in the omelette. The egg mixture must not be seared or allowed to turn brown and the individual layers should stick together. The Japanese use a small rectangular pan so that the *tamago* comes out box-shaped, but you can also use an ordinary round pan and trim the edges before using the omelette. Remove the omelette from the pan and press it into shape with a bamboo rolling mat, which will imprint a nice surface texture on the *tamago*. When the omelette is sliced crosswise, the *nori* will stand out against the yellow of the omelette as fine black folded lines.

Seaweeds in bread, pasta, and savory tarts

In former times in the Nordic countries, when supplies in the pantry were beginning to run low in the winter, both brown and red algae were used to make the flour used for baking bread go a little further.

In the modern Western countries, where seaweeds as food are relatively underappreciated, incorporating seaweed flour into bread dough is possibly the most practical way to entice people into enjoying the wealth of vitamins, proteins, healthy fats, and minerals that are to be found in the sea. Seaweeds can replace salt in bread, pizza dough, and pasta. It can also be substituted for some of the refined flour that is the source of the white carbohydrates that are far too dominant in the Western diet.

Seaweed meal can actually be purchased in many countries, but it is often designated as animal fodder, for example, for horses, ruminants, and fish. In Japan, on the other hand, it is an ingredient in noodles and different types of cakes. Seaweed meal can be used both as granules or ground into flour. For pasta, of course, a very finely ground flour is needed, otherwise the pasta dough will not stick together properly. Typically, one can replace 1–10% of wheat flour by seaweed flour, depending on the type of dough in question. Experimentation has shown that adding this amount of seaweed flour does not really alter the physical properties of bread dough, nor does it inhibit its ability to rise. The taste and aroma of the seaweeds are preserved. Kelp flour can be too salty, but other types, for example, those made from winged kelp, are quite suitable. Importantly, seaweeds can replace some of the table salt in bread.

It has been suggested that mixing seaweed granules with white flour should help to encapsulate the starch in the flour and result in a slower release of the nutritional elements of the bread in the digestive system.

Not all types of seaweeds are equally suitable for adding to bread dough. Some of the best types are dulse, sea lettuce, purple laver, winged kelp, bullwhip kelp, and giant kelp, whereas sugar kelp, oarweed, and tangleweed are not good.

RYE BREAD WITH SEAWEEDS

It is easy to play around with adding seaweed flour to bread and pasta, until one finds a combination that suits one's personal taste. Try to substitute seaweed flour for 1–5% of the regular flour in your favorite recipe. Seaweed flour from the milder tasting varieties can replace white and rye flour to a greater extent than that from kelp. This is my wife's recipe for rye bread with seaweed flour.

Rye bread with seaweeds   *makes 2 loaves*

2 tbsp seaweed flour or granulated kelp
sourdough starter
rye flour
white flour
2 tbsp salt
1 package of mixed seeds
1 bottle low-alcohol beer

Dissolve the sourdough starter in 1 liter water. Add rye flour and white flour in equal quantities and mix to form a soft mash. Let the mash rest overnight. Then mix in the salt, beer, seeds, seaweed flour or granules, and enough rye flour and white flour, again in equal quantities, to make a stiff mash. Set aside some of the dough and keep it cold for use as a starter next time. Put the dough in greased loaf tins. Bake 1 hour at 100°C and 1 hour at 200°C.

BUNS AND CRISPBREAD WITH SEAWEEDS

Buns and cripbreads with the salty and briny taste of the ocean can be made by adding seaweed granules or pieces to the dough. The buns are especially good served with a fish dish or soups and the crispbreads with cheese.

Kirsten's flaxseed buns with seaweeds   *makes about 20 buns*

2 tbsp dried seaweeds (dulse, winged kelp, sea lettuce), finely chopped

50g cake yeast (or approx. 20 grams dry yeast)

400ml lukewarm water

7 tbsp flaxseed

2 tsp honey

200ml plain drinking yogurt

3 tbsp grapeseed oil

2 tsp salt

400g coarse flour (graham, spelt, whole wheat)

400g white flour

1 egg, lightly beaten

Mix the flaxseed and seaweeds with 300ml lukewarm water (ca. 40°C) and let stand to plump for about 15 minutes. Mix the cake yeast with 100ml lukewarm water (ca. 37°C). If using dry yeast, follow directions on the package and adjust the volume of water used to plump the seaweeds accordingly. Mix all the wet ingredients and the salt together. Knead in the flour a little at a time; the dough is ready when it no longer sticks to the hands but is still elastic. Set the dough aside to rise for about an hour. Punch down the dough and turn it out onto a floured surface. Shape the buns and place them on a greased baking sheet. Allow the buns to rise for about 15 minutes. Brush them with beaten egg and bake for ca. 20 minutes at 200°C.

▶ Fine and coarsely chopped *mekabu* from *wakame* sporophylls.

166

Julie's crispbread with seaweeds *makes about 50 pieces*

150 ml rolled oats

150 ml flaxseed

100 ml sunflower seed

100 ml pumpkin seed

2 tsp salt

250 ml flour

1 tsp baking powder

4 tbsp mixed seaweed granules (sea lettuce, dulse, bullwhip kelp,
    giant kelp, *mekabu*)

200 ml water

2 tbsp grapeseed oil

Seaweeds in bread, pasta, and savory tarts

In a bowl mix together the oats, seeds, seaweeds, salt, and baking powder. Add water and mix well until the dough becomes sticky. Divide the dough into two and place one part on a piece of baking paper. On top of the dough add another a piece of baking paper and roll the dough out evenly and as thinly as possible between the two. With a knife or pizza wheel cut the top baking paper and divide the dough into squares without cutting through the bottom paper. Remove the top baking paper and place the dough and the bottom paper on a baking sheet. Repeat the procedure with the other part of the dough. Bake the crispbread at 200°C for about 15–20 minutes until the bread is golden brown. Let the crispbread cool on a baking rack. After a few minutes the crispbread can be broken along the scored lines.

◀ Crispbread with seaweeds.

QUICHE WITH SEAWEEDS

I was given this recipe by Rae Hopkins. It is almost identical to one for an ordinary spinach quiche.

Rae's seaweed quiche   *serves 6*

Crust:

5g small seaweed flakes (sea tangle, *wakame*, winged kelp, or
 bullwhip kelp) or thin strips of *mekabu*
375ml coarse wheat flour, 375ml wheat germ
1 tsp salt
125ml margarine
125–150ml cold water

In a bowl mix together the seaweed flakes, flour, salt, and wheat germ. Cut in the margarine. Add water and mix in until the dough sticks together. Roll out the dough and place it in a greased pie plate. The strips of *mekabu* add flavor to the crust and may help keep it together.

Filling:

125ml dried *wakame* or winged kelp, cut in strips
125ml spring onion, chopped
250ml mushrooms, sliced
250ml cheddar cheese, grated
3 eggs, lightly beaten
250ml light cream
125ml milk
1 tsp prepared mustard
cayenne pepper
2 tbsp butter or margarine

Soak the seaweeds in water and discard the water. Mix the seaweeds with the onions, mushrooms, and cheese and place on the crust. Stir together the eggs, cream, milk, mustard, and cayenne pepper and pour on top. Dot the filling with small pieces of butter or margarine. Make sure that the seaweeds are covered with liquid so that they do not become hard during baking. Bake at 180°C for about 45 minutes.

Seaweed pesto on bread and crackers

The preparation method for laverbread (above) can easily be adapted for use with other species. They can be puréed together with a variety of other taste substances to make a sort of pesto. This seaweed pesto is served cold on bread or crackers as a snack or canapé.

Seaweed pesto

200ml rehydrated seaweeds (*wakame*, sea lettuce, winged kelp, giant kelp)

400ml mixed greens, for example, parsley, purslane, spinach, or basil

100ml pine nuts

3 cloves garlic

2 tbsp parmesan cheese

2 tbsp lemon juice

100ml olive oil

Blend the solid ingredients in a food processor, adding olive oil and lemon juice gradually until the desired consistency is achieved. Serve on fresh baked bread or crackers and, possibly, with some cheese. The seaweed pesto keeps for 3–4 days in the refrigerator.

Pizza with seaweeds

Seaweeds can easily be added to homemade pizza. Granulated, powdered, or ground seaweeds can be added to the pizza dough, which then does not need any additional salt. One can also mix dried flakes of dulse or sea lettuce into the dough. Rehydrated winged kelp, *wakame*, and *konbu* can be mixed into the sauce that goes on top of the pizza and a seaweed salad can be substituted for the arugula salad often served with it.

Seaweeds and cheese

The salty, slightly smoky, and iodine-like flavors and *umami* taste of brown algae are a natural complement to cheeses. When paired with cheeses like parmesan and emmental, which already have characteristic *umami* tastes and nutty flavors of their own, the effect is greatly enhanced. You can make an easy appetizer by wrapping a strip of softened *wakame*, giant kelp, or bullwhip kelp around a slightly firmer cheese, such as a large cube of smoked mozzarella or a disc of goat cheese. I have become very enthusiastic about the combination of seaweeds and soft and lightly smoked, unripened cheese, a variety of which is a specialty of my home town.

Soft unripened cheese wrapped in seaweed   *serves 4*

4 blades of dried seaweed (*wakame*, giant kelp, or bullwhip kelp),
    ca. 15 cm × 3 cm
500 g soft unripened cheese, in one piece

Soak the seaweed blades in lukewarm water and drain them on a piece of paper towelling. Using a large cookie cutter or a sharp knife, cut the cheese into pieces with a decorative regular shape—e.g., circle, square, or triangle. Be sure to moisten the cutter or the knife with lots of water so that the cheese does not stick to it. Wrap a seaweed blade, cut to the right size, around each piece. Serve as an appetizer, possibly with a little green *wakame* salad. To add yet another subtle taste, sprinkle a little *nori-furikake* on top of the cheese just before serving.

To make an easy appetizer, fold softened *wakame*, giant kelp, or bullwhip kelp around cubes or small rolls of firm unripened cheese, such as goat cheese, or cream cheese flavored with seaweed ash salt and chili.

Dulse tastes equally good with unripened cheeses. Here is a simple recipe which can be made with dulse, *wakame*, or winged kelp.

Cream cheese with dulse

1 tbsp dulse or smoked dulse, softened and chopped finely
200 g cream cheese

Thoroughly mix the seaweed bits into the cream cheese and serve on coarse oatcakes or spelt crackers.

Maine Coast Sea Vegetables markets dried dulse that is lightly smoked over applewood. It is great with cheese, in particular lightly smoked, unripened cheese or a mild brie. To enhance the flavor, drink a glass of single malt, smoky whisky with the cheese.

Brie with applewood smoked dulse _____

mild brie cut into thick strips
smoked dulse

Toast the dried dulse so that it is totally crisp and then crush it into coarse granules. Roll the pieces of brie in the granules to cover their surfaces. Serve as a finger food with a glass of wine or a beer.

Seaweeds in desserts and cakes

In desserts, seaweeds make an appearance almost exclusively as gelling agents, especially as agar. Here it is so commonly used that it is surprising that only a few people know that they are actually eating a seaweed product. Candied seaweed is found in a type of desserts that we might, instead, think of as a snack because they are rarely very sweet. But seaweeds, in their own right, can easily be used in a variety of ways in desserts and cakes, just as they are in savory dishes.

Seaweeds with fresh fruit

Flakes of toasted, crushed winged kelp have such a mild taste that they can be sprinkled on a fruit dessert or on slices of fresh melon, apples, pears, oranges, or bananas. Their slightly salty and nutty taste nuances provide an interesting contrast to the sweet and tart taste of the fruits.

Seaweeds as gelling agents in desserts

Agar and carrageenan, polysaccharides derived from red algae, have thickening and stabilizing properties and have been widely used as food additives for generations.

Agar is a popular gelling agent for desserts. Its advantages are that it sets liquids firmly and stably and, in addition, is tasteless and odorless. The drawback in using it is that its melting point is 85°C. Consequently, desserts made with it have a different mouthfeel from those thickened with pectin or gelatine, which melt or have a soft feeling in the mouth. On the other hand, agar, especially in powder form, is often much easier to work with.

Certain fruits, such as papaya and pineapple, contain enzymes that break down gelatine. The polysaccharides in agar and carrageenan, however, are not affected by these enzymes and they can be used as gelling agents for desserts containing these fruits. Agar does not work well in an acidic environment, so it is best to use carrageenan for desserts that involve fruit juice, coffee, or oxalic acid from rhubarb.

Agar is sold as a powder or in flakes—note that the volume of flakes is about 50% greater than that of powder. It has to be soaked in cold water for about 10 minutes and then boiled for at least 10 minutes until the agar is completely dissolved. As the mixture will set when it cools to below ca. 38°C, only those ingredients that tolerate being cooked should be added while the agar is being boiled. The agar mixture can then be added to the remaining ingredients while it is still at a temperature above 38°C, stirring constantly so that it does not form lumps.

BLANCMANGE—A VERY BASIC PUDDING

The most classical dessert that is stiffened with agar is blancmange. Here is a recipe for a version that is almond flavored.

Almond pudding *serves 4*

2 tbsp agar (powder)
125g blanched almonds
125ml milk
125ml water
125ml sugar
250ml whipping cream
2 tsp acacia or maple syrup

First prepare the almond milk by crushing the almonds with a mortar and pestle or in a blender. Add the crushed almonds to the water and milk, stirring the mixture thoroughly to release the flavor substances from the almonds. Sieve the mixture through a cloth. This is the almond milk. Place the cream, sugar, and agar in a bowl in a water bath and warm it to the boiling point. Stir the mixture vigorously to dissolve the agar and then add the almond milk and the syrup. Pour into a cold pudding mold and cool for several hours until the liquid has set completely. To serve, unmold onto a plate and garnish with fresh fruit, if desired. To make a fruit-flavored pudding, replace some of the liquid with fruit juice.

PUDDING WITH CARRAGEEN

There are many traditional recipes for desserts that can be gelled with carrageenan from carrageen (*Chondrus crispus*). Because it is especially good at binding proteins, carrageenan is an ideal gelling agent for puddings that contain milk, cream, and yogurt.

Vanilla pudding with carrageen
serves 4

10g dried carrageen
500ml milk
1/8 tsp sea salt
1–2 tsp vanilla sugar

Soak the seaweeds in cold water for a half hour and then wash them carefully to remove any small shells that sometimes cling to them. Squeeze the seaweeds, tie them inside a piece of cheesecloth, and place them in a bowl together with the milk. Put the bowl in a water bath and allow the mixture to simmer for about half an hour. From time to time press down on the bag with a spoon to release the carrageenan from the seaweeds into the milk. Remove the bowl from the water bath, discard the bag, and add the salt and vanilla sugar. Stir vigorously. Transfer the mixture to a cold pudding mold and allow it to cool for a few hours until the pudding is completely set. To serve, unmold onto a plate and garnish with fresh fruit or a cherry sauce. Note that the pudding may take on a slightly greyish color from the seaweeds.

YŌKAN—JAPANESE DESSERT JELLY

Many Japanese desserts and confections are based on a paste made from small, sweet *azuki* beans. As beans from the red variety are sweeter than those from the green one, they can be used to make some of the healthiest desserts imaginable. A stiff jelly, known as *yōkan*, is produced from *azuki* bean paste and eaten like a candy. Because it is set with agar, it has a very solid, chewy mouthfeel. Often *yōkan* is mixed with powdered green tea, *maccha*, and eaten as a jelly or used as a filling in cakes and confections. These have a limited shelf life and must be consumed when very fresh. The bean paste can be made from dried *azuki* beans or bought ready-made in cans in specialty stores. Sometimes whole, but soft, beans are included in the paste and it is pleasant to chew on these.

Yōkan red bean dessert jelly   *serves 4*

1 tbsp agar

50 g paste from red *azuki* beans

200 g sugar

200 ml water

2 tsp *maccha* (optional)

Stir the agar into 150 ml water and cook for about 10 minutes, until the agar is completely dissolved. In a pot mix together the bean paste, the rest of the water, sugar, and *maccha*, if desired, and bring it all to a boil. Note that the *maccha* must first be whisked into a little water so that it does not form lumps. Stir in the agar mixture, allow to cool slightly, and pour into a metal pan. The dessert is ready when the jelly is completely set. To release the jelly from the pan, put the pan in a water bath with cooking water for a short time to loosen the jelly around the edges. Turn the jelly out carefully onto a plate. For a more colorful effect, make one portion of jelly with *maccha* and one without. Pour the first portion into the mold, allow it to set, and then add the second portion. The uncut jelly is served as is and sliced at the table.

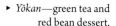

▸ *Yōkan*—green tea and red bean dessert.

LEMON CAKE WITH SEA LETTUCE

Milder varieties of seaweeds, such as dulse and sea lettuce, can be used for baking, both in breads and in cakes. The seaweeds impart a fine-tuned salty taste and at the same time add a colorful touch—red from dulse and green from sea lettuce.

Lemon cake with sea lettuce   *serves 4*

2 tbsp flakes of dried sea lettuce or dulse

120 g white flour

120 g butter

2 eggs

120g icing (or confectioner's) sugar

2 organic lemons

2 tsp baking powder

sugar

Grate the zest from one of the lemons and mix it with the butter which has been melted with the icing sugar. Add the seaweed flakes, flour, egg yolks, and baking powder and mix everything together. Beat the egg whites and fold them into the mixture. Pour into a greased ring pan and bake for 20–25 minutes at 200°C. Squeeze the juice from the lemons and bring to a boil with sugar to taste. When the cake is baked, turn it out of the pan and drizzle the sweetened lemon juice over it.

ICE CREAM WITH SIMMERED KELP

Some of the large brown seaweeds, kelp such as *konbu* and oarweed, are good condiments with ice cream because, after being simmered, they remain soft and tender even at low temperatures. Sugar kelp is less useful because it excretes too much slimy polysaccharide. Simmered kelp can be purchased in Japanese stores under the name *konbu tsukudani*. If not available, simmered kelp is easily made by soaking dry kelp in water for an hour until it is fully rehydrated and then shredding it into fine strips. The strips are simmered for at least half an hour in a mixture of soy sauce and *mirin* (or sweet white wine) together with finely grated *shiitake*. Simmer until all the liquid is caramellized and the kelp is glossy and sticky. The final product, which is a genuine *umami* flavor bomb, is mixed into a softened ice cream that has a neutral flavor. Roasted white sesame seeds can also be added to the mixture.

Seaweeds in desserts and cakes

▲ Sea lettuce.

◄ Ice cream with simmered kelp (*konbu tsukudani*).

Seaweeds in drinks

Seaweed tea

Dried bladder wrack has been used for ages to make a type of herbal tea, which, according to folk medicine, is an effective headache remedy. Nowadays, bladder wrack is found in some drinks sold in health food stores. A tea can be brewed from just about any variety of dried seaweed, for example, dulse, bladder wrack, and sea tangle.

When dried seaweeds are soaked in warm water, minerals, salts, soluble carbohydrates, and iodine are extracted, giving the water a strong taste that some find agreeable. The infusion can be drunk warm or cooled. Optionally, green tea leaves, grated ginger, lemon juice, or honey can be added to the infusion.

To make the extract from dried bladder wrack, boil it in water for a quarter of an hour or pour boiling water over it and let it steep overnight. The fastest way to make it is to whisk powdered bladder wrack into boiling water.

As far as I am concerned, the tastiest seaweed tea is one that can be made very simply from salty-sweet, *konbu tsukudani*, a common Japanese breakfast food. This type of *konbu* is marinated with vinegar, *shiitake*, and sweetened soy sauce or *mirin* and then simmered for a long time, which makes the *konbu* very soft. The marinade mixes with the polysaccharides exuded by the seaweed. The tea is dark brown and has a strong taste of the sea, iodine, and a little licorice.

Another variation of *konbu* tea has a fruity plum or apricot flavor from pickled plums, *umeboshi*.

Konbu tea   *makes 1 cup*

Place a couple of small pieces, about 2cm×2cm, of marinated *konbu* (*konbu tsukudani*) in a tea cup and pour boiling water over them. Allow to steep for less than a minute. The *konbu* pieces can be eaten as the tea is drunk.

Nereo tea with *umeboshi*   *makes 1 cup*

Place a few pieces, according to taste, of dried bullwhip kelp and 1–2 small *umeboshi* (brine-pickled Japanese apricots) in a tea pot with 800ml boiling water. Allow to steep for 5 minutes and sweeten with honey to taste.

▲ Scottish seaweed beer from the Williams Brothers Brewing Company.

SEAWEEDS IN WINE, BEER, AND SPIRITS

Seaweeds contain only a small quantity of carbohydrates that can be fermented with yeast. In order to produce wine or beer with an appropriate level of alcohol it is necessary to add sugar or honey. Consequently, the alcohol content is mostly derived from the fermentation of these sugars. But the taste comes from the seaweeds and brown, red, and green varieties have all been used for this purpose. The finished product retains a large proportion of the nutritious substances from the seaweeds, including the vitamins and minerals.

A German marine biologist has made a wine that has playfully been dubbed 'Chateau Seaweed'. It has a crisp slightly salty taste with overtones of sherry and marzipan. This fortified wine is made from sugar kelp and has an alcohol content of 16%. Research has shown that the antioxidant properties of the seaweeds break down in short order when the wine is aged.

In Scotland, the Williams Brothers Brewing Company adds small quantities of bladder wrack to the mash to make Kelpie Seaweed Ale, a rich, dark beer with an alcohol content of about 4.5%. It is lightly salty and has a strong taste of malt, a little chocolate, and roasted barley, but no discernible traces of seaweeds or iodine. Personally, I find more it interesting than flavorful.

Beer and wine makers are probably most familiar with seaweeds as fining agents. Carrageen is very suitable for fining homebrewed wine and beer, which contains proteins in the form of tiny particles that cause cloudiness. The carrageenan content of the seaweeds binds with these proteins, clumping them together so that they fall to the bottom.

It is a tradition in Japan to make strong liquors called *shochu* flavored with a 5% extract of the brown alga, *konbu*. Recently, the world famous Danish Restaurant noma, which is the leading exponent of the so-called New Nordic Cuisine, introduced a seaweed-flavored aqua vitae. Its special taste is derived from an extract of Icelandic dulse, which turns out to impart *umami* and a particular floral flavor to the drink.

Some of my very good friends, who know of my addiction to seaweeds, recently presented me with a very special bottle of single malt Islay whisky called Celp, The Seaweed Experience. This unique fluid is produced in Scotland by The Ultimate Whisky Company and distilled at Lagavulin. The color is a distinctive green, enhanced by the color of some unspecified seaweeds floating inside the bottle. The taste is very salty, peaty, and an ultimate experience of the sea.

Seaweed flavor notes might just be the new frontier in bartending.

Seaweeds in drinks

▲ Left: German wine made from sugar kelp. Right: A Danish aqua vitae based on an extract of Icelandic dulse.

▲ Leftt: Japanese hard liquor, *konbu-shochu*, made with a 5% extract of *konbu*. Right: A Scottish single malt Islay whisky flavored with seaweeds.

Spirulina smoothie

The blue-green microalga Spirulina has gained great currency as a veritable protein bomb in fruit smoothies. Apart from its nutritional qualities, Spirulina imparts a beautiful, bright green color to the drink.

▲ Fruit smoothie with Spirulina.

▶ Spirulina as freeze-dried powder and as granules.

Julie's fruit smoothie with Spirulina	*makes 2 large glasses*

1 tbsp Spirulina
100ml orange or tropical fruit juice
200g fresh, ripe fruit, for example, pineapple or mango
1 ripe banana
juice from half a lime
6 ice cubes

Blend together all the ingredients. For a thicker smoothie, use more fruit or less juice.

Seaweed wellness drinks

Minerals, salts, vitamins, polysaccharides, and various flavorful compounds, e.g., free amino acids like glutamate that imparts *umami* flavor to *dashi*, can be extracted from seaweeds by soaking them in water. The resulting tastes can vary, depending on whether the algae are steeped in warm or in cold water. Seaweeds, such as sugar kelp, that exude large quantities of polysaccharides are not well suited for making mineral drinks. Dulse and some of the finer brown seaweeds, such as winged kelp, can be used as a basis for drinks that are not overpowered by the taste of the sea.

The small Danish company Jens Møller Products, which invented Cavi-art, described earlier in this chapter, has launched what is possibly the world's first series of drinks based on brown seaweeds, under the trade mark VITA-ALL. The drinks, which are are slightly sweetened with sugar and mixed with various fruit juices, are marketed as a healthy alternative to soft drinks.

▶▶ VITA-ALL. Seaweed wellness drink with and without black currant.

Seaweeds in snacks

SNACKS

When one is longing for something salty or a combination of sweet and salty, toasted seaweed chips are a good choice, not only because they have an abundance of potassium salts, but because they taste good. In the mouth, the dried seaweeds will try to recover the water they lost during drying and toasting. Therefore, drink plenty of liquid with them, both to slake the thirst brought on by the salt and to soften the dietary fiber in the seaweed. In the early 1900's, toasted dulse was served as a snack in Irish and Scottish pubs to stimulate beer consumption. On the Westman Islands and other parts of Iceland, dried dulse is still a common snack food.

The simplest way to make seaweed chips is to toast the pieces of seaweed on a flat toaster or on a very hot pan. The red algae dulse and laver and the brown algae *wakame*, bullwhip kelp, and giant kelp are the species that work best. The seaweed should be toasted only until it has just become crisp and must not be allowed to burn. As it toasts, the seaweed changes color from dark green or red to take on more yellow and brown tones.

Snacks made from some of the milder varieties of seaweeds can easily be considered as a sort of candy, which is healthy to boot.

CHIPS

Chips are very easy to make from the more delicate types of kelp.

> ### Kelp chips
> dried kelp, e.g., winged kelp, *wakame*, or bullwhip kelp
> oil for frying, preferably canola oil
>
> With a knife or scissors cut the seaweed into bite-sized pieces and place them in a pot with hot oil. After a very short time, the seaweed pieces will turn yellow and start to puff up. Remove them with a slotted spoon and pat them dry on a piece of paper towelling. These wonderfully crisp chips can also be crushed and sprinkled on vegetables, salads, fish dishes, and soups.

The spore-bearing blades, sporophylls, on winged kelp and *wakame* are truly delicious because they are slightly sweet and contain two to three times as

▶▶ Dried giant kelp.

much fat as the other blades of the seaweed. Sporophylls can be deep-fried or candied and coated with sesame seeds. *Wakame* sporophylls, called *mekabu* in Japan, are especially rich in fucoidan, which is thought to have anti-cancer properties.

Toasted strips of *nori*, about 3cm×7cm, seasoned with soy sauce or other spices, can be bought in small packages and eaten as a snack. Dried *nori* can be heated on a hot pan, so that it puffs up like popcorn.

▸ Sesame seaweed chips made with candied *wakame* sporophylls.

RICE CAKES WITH *NORI*

Senbai, baked or grilled rice cakes, are among the most common Japanese snack food. They are salty-sweet biscuits that can be large or small, come in various shapes, and display many regional variations. The rice cakes can be very lightly spiced with soy sauce and *mirin*. They often have a piece of *nori* stuck on top or wrapped around them. *Senbai* are made from *mochi*, a soft, Japanese rice cake made from steamed white rice that is mashed into a flexible paste, which is then baked or roasted over a charcoal grill.

Senbai are very crisp and, thanks to the *nori*, have a taste of the sea. They can be eaten as a sort of candy and are often served with a cup of green tea.

▸ *Senbai*—Japanese rice cakes with *nori*.

Seaweeds and chocolate

Some of the taste substances in seaweeds, notably those with overtones of salt, nuts, and iodine, are perfect partners for the bitter sweetness of chocolate. Roasted and crisp seaweed blades of bullwhip kelp or giant kelp can be dipped in dark chocolate and eaten as a snack or with an ice cream dessert.

Seaweeds in snacks

◄ Bullwhip kelp coated with chocolate.

The ever-versatile dulse, with its slightly salty and nutty taste, also combines well with dark chocolate that has a high cocoa content. Lightly smoked dulse makes the flavor experience even greater.

◄ Applewood smoked dulse on dark chocolate buttons.

Although such confections are hard to find, there are a few commercial chocolate products available that consist of chocolate slabs with dried sea lettuce and dulse flakes sprinkled on the back.

◄ Commercial chocolate bar with dulse and sea lettuce.

HERRING ROE ON KELP

Herring spawn onto stones or on a sandy seabed, but also onto seaweed blades. Herring roe on kelp (*komochi wakame*) is a great delicacy in Japan. In Alaska, 15% of the commercial herring roe production is derived from roe that is spawned onto kelp. The fish deposit a layer of eggs, which can be up to two centimeters thick, on the seaweed blades, especially those of the large species such as giant kelp (*Macrocystis pyrifera*). The blades are a perfect substrate for the slimy roe, which can be deposited either onto one side or, more often, both sides of the blades. Commercial exploitation of herring roe in this way is more sustainable than the usual method, which is to catch herring as they are about to spawn and cutting the fish open to remove the egg sacs.

Roe-bearing seaweed blades are normally cut into strips and placed in layers of salt or frozen. After they have been thoroughly soaked in water, the strips can be eaten as a snack or served as an appetizer. The slightly chewy, but soft, texture of the seaweeds complements the crunchy crispness and salty taste of the fish roe and brine.

Herring roe on kelp is a traditional food of the first inhabitants of the west coast of Canada. The roe was a highly prized part of a festive meal, just as it was an important medium of exchange. The native peoples harvested large seaweed blades and anchored them with rocks on the seabed close to the mouth of a river to increase the likelihood that the herring would spawn on them. The roe-bearing blades were preserved by drying them on rocks in the sun. They could then be bundled up and transported for trading purposes. Cut into strips, they were eaten as a snack or as candy. The dried roe could also easily be made part of a meal by soaking it in water or frying it in fat.

Seaweeds in the kitchen

▸ Herring roe on giant kelp (*Macrocystis pyrifera*).

Seaweeds in gastronomy

SEAWEED INNOVATIONS AND *HAUTE CUISINE*

In the foregoing part of this chapter I have tried to demonstrate the many simple ways in which seaweeds can be incorporated into everyday cooking, using recipes that anyone can follow to enhance and add flavor to a variety of familiar dishes. Almost without exception, the underlying idea is to encourage the reader to introduce seaweeds incrementally into the everyday diet and to pave the way, over time, for his or her own discoveries of how to reap the nutritional benefits they contain.

As we set out to do so, we can turn for inspiration to the greater prominence that has been given to marine algae during the last few years by many serious practitioners of *haute cuisine*. This has, to a certain extent, been driven by a felicitous combination of factors: the locavore movement, a heightened consciousness of the importance of sustainable food sources, the quest for *terroir*, and the desirability of eating healthy, pesticide-free foods. While the ways in which innovative chefs use seaweeds in elaborate gourmet meals probably lie well beyond our capacity to reproduce them for ourselves at home, it is worth delving a little deeper into what happens behind the scenes in the kitchens of some of the finest restaurants.

NEW RAW MATERIALS OPEN NEW GASTRONOMIC PATHWAYS

For the professional chef, the challenge of how to use raw materials is on a completely different plane from the one we face at home. Like art and the search for new knowledge, gastronomy is driven by a combination of craft, desire for renewal, self-criticism, vision, and delight in playing with new ingredients and ideas. Here, the goal is to invent fabulous new dishes, where seaweeds have been incorporated as essential, and often surprising, elements, which elevate the resulting creation to something unique and, possibly, sublime. Something that calls out for both sensory and esthetic appreciation.

For most chefs in Europe and North America seaweeds are a relative novelty. This challenges them to rework traditional recipes in daring ways and to devise pioneering gastronomic experiences. Some have drawn inspiration from classical Asian cuisine, where seaweeds have never fallen out of favor. Other well-known chefs have embraced the use of seaweeds in their kitchens and restaurants in order to capitalize on the profusion of tastes, textures, shapes, and colors that are characteristic of marine algae. Often this involves making excursions to harvest fresh seaweeds from their local surroundings, a natural extension of foraging and the quest for wild, edible plants and herbs in forests and fields that gained momentum a few years ago.

You might wonder about how the patrons of these high-end, expensive restaurants react to finding seaweeds on the menu. It seems that part of the appeal lies in the ingredients themselves. The diners appear to be intrigued by the idea of eating seaweeds and take an interest in learning more about them and their places of origin. In this way, tasty marine algae may win a permanent place on these menus and, in the future, be featured more prominently—a vindication for the food of our ancestors.

FUN AND GAMES IN THE KITCHEN—WHEN 'PLAY' BECOMES MORE SERIOUS

For the majority of innovative chefs, preparing food and creating new gastronomic experiences has always been a question of playing with the ingredients and giving free rein to imagination and intuition. But now it has become more quantitative and the experimentation often takes place in settings devoted to molecular gastronomy that are much more like highly disciplined university science laboratories than the somewhat hectic work areas of commercial kitchens.

René Redzepi, the owner of noma, the world famous Copenhagen restaurant, along with many who come to learn from him, carries out precise research in the nearby Nordic Food Lab. In Britain at The Fat Duck, Heston Blumenthal has set up a research and development kitchen and uses scientifically based approaches in his cooking. Much farther to the south in Spain, another stellar restaurant, elBulli, recently closed and might be converted into a 'gastronomic think-tank', dedicated to elevating avant-garde cuisine to even greater heights. In these places, new techniques are constantly evolving and being combined with those already tested in other contexts. Like all true research, the work starts out with no preconceived notions—new recipes are the products of bold and persistent experimentation. Part of the appeal offered by incorporating seaweeds into these dishes lies in their texture and their capacity both to season and to absorb tastes from other food ingredients.

TAKING HOME COOKING TO NEW LEVELS

▸ ▸ Samples of seaweeds are tested for color and flavor. Chefs having fun and playing in the kitchen.

Once you have become comfortable with using seaweeds to augment traditional dishes and your own cooking, you might find inspiration in the following descriptions of how professional chefs create high-end gastronomical delights.

186

Veal tartare with Harry's *crème* and dulse

Chef Klavs Styrbæk, owner and chef at Kvægtorvet (The Cattle Market), located in Odense, Denmark, is known for striking a delicate balance between innovation and tradition with a keen eye for high-quality products from his local area. His classic book 'Grandmother's Kitchen', as well as his book 'Umami' written in collaboration with the present author, is steeped in his vision for the revival of the traditional Danish cuisine.

Klavs has recently experimented with the use of local seaweeds, as he is particularly interested in using them to impart *umami* flavors to his dishes. For example, he marinates bladder wrack in rapeseed oil or smokes it over wood from old fruit trees that grow in his own garden.

In this recipe, Klavs adds *umami* to a raw meat dish by using Icelandic dulse and dresses it with a sauce that echoes the one for carpaccio at Harry's Bar in Venice.

- *Harry's crème:* All ingredients must be at room temperature. Prepare a mayonnaise by whisking together one egg yolk, ½ tsp mustard, a few drops of lemon juice, 1 tsp white wine vinegar, salt, and freshly ground black pepper. Add 2½ dl olive oil or rapeseed oil, a few drops at a time at first and then in a thin stream, until the mixture has the desired consistency. Adjust the seasoning with a reduction of Worcestershire sauce, 2 tbsp cream, and possibly a few drops of Tabasco sauce and a little more lemon juice. The sauce should be thick, but not firm, and slightly piquant.
- Process a large piece of fresh ginger in a juicer to produce 2 tbsp juice.
- Mince 300g of veal into tartare and mix in enough ginger juice to give a somewhat sharp taste. Add rapeseed oil, salt, freshly ground pepper, and mix until the taste is soft and spicy. Shape into two equal patties.
- Peel 20 fresh shrimp and marinate them in horseradish juice or finely ground horseradish mixed with a bit of salt.
- Shell broad beans or fresh peas, blanch them for 5 seconds in boiling water, and then plunge them briefly into ice water.
- Cut a shallot into thin slices and put the slices in ice water until needed.
- *To serve:* Spread a thin layer of Harry's *crème* on a plate and put the tartare on top of it. Place the marinated shrimp on top of the tartare and dress with beans, dulse, scallion slices, and a few pieces of watercress.

▶▶ Veal tartare with Harry's *crème* and dulse.

Fresh shrimp with pickled sea lettuce and beach herbs   *serves 4*

Chef René Redzepi, a pioneer of the New Nordic Cuisine, is very preoccupied with finding regional raw materials for his kitchen at Restaurant noma in Copenhagen. He expresses his philosophy as follows: "In an effort to shape our way of cooking, we look to our landscape and delve into our ingredients and culture, hoping to rediscover our history and shape our future." The restaurant has been spearheading the rediscovery of *terroir* in the Nordic cuisine. Lately, René has developed an interest in using seaweeds from different places in the northern countries. He thinks that seaweeds will come to play a major role in the gastronomy of the future and that there is a real need to make them available locally.

In this recipe, René combines ingredients from three biological kingdoms: shellfish (shrimp) from the ocean, seaweed (sea lettuce) from the waters of the coastline, and herbs (stonecrop) from the beach.

- Carefully remove the shells from 20 fresh deep-water shrimp starting at the tail end and working towards the head. Line the shrimp up on a plate and keep them cool on ice.
- Rinse 50g of fresh sea lettuce very thoroughly to free it from sand. Cut baking paper into 4 circles, 13cm in diameter. Pickle the sea lettuce by rinsing it again in a mixture of 45ml vinegar and 90ml water. Spread it out on the baking paper circles, making sure it fully covers each surface without being stacked in several layers.
- Peel one rhubarb stem and dice finely. Cook the pieces *sous vide* for 6 minutes with 50g of a stock syrup made of 50% water and 50% sugar. Strain the rhubarb, mix the pieces with grapeseed oil, and season with salt before serving.
- Make a rhubarb juice by mixing 110g rhubarb juice, 80g beet juice, 50g balsamic apple vinegar, and 50g stock syrup. Strain the juice.
- Put 16 small stonecrop leaves, 24 small sea purslane leaves, and 24 small beach mustard leaves in ice water and then spin them dry. Keep them in the refrigerator until needed.
- *To serve:* Take cold plates from the refrigerator in the last instant before serving. Season the shrimp, arrange on plates, and cover with the sheets of pickled seaweed. Dress with the diced rhubarb, the beach herbs, and a few spoonfuls of the juice.

▸ ▸ Fresh shrimp with pickled sea lettuce and beach herbs.

Ice cream with dulse

Chef Lars Williams trained at several of the world's top restaurants including wd~50 in New York, The Fat Duck in England, and Restaurant noma in Copenhagen. He has been Head of Research and Development at the Nordic Food Lab, a non-profit self-governed institution established by the head chef of noma, René Redzepi, and gastronomic entrepreneur Claus Meyer, with the purpose of exploring Nordic Cuisine, cornerstones of gastronomy, and disseminating the results of this exploration. Lars works as a hybrid between a chef and a scientist. Among other things, he has studied the use of Nordic seaweeds in cooking and gastronomy. His current research interests include fermentation and new *umami* flavors.

In this recipe, Lars uses a flavorful extract from the red seaweed dulse to infuse an ice cream in order to impart both *umami* and floral notes to the ice cream.

- Place 12g dry dulse in 600ml milk in a plastic bag under vacuum and seal, leaving it in a refrigerator overnight to cold-infuse.
- Strain the dulse, blend it into a fine purée, and reserve it for later use.
- Dissolve 35g sugar and 80g trimoline (inverted sugar syrup) in a small amount of warmed milk.
- When this mixture has cooled, add to it the rest of the milk and the dulse purée, together with 100ml cream and 24g ColdSwell cornstrach,. Mix thoroughly, and freeze the mixture in Pacojet containers.
- Just before serving, prepare the ice cream in the Pacojet by high-speed precision spinning and thin-layer shaving to give it a creamy consistency.

The dulse ice cream was conceived to demonstrate the culinary versatility of seaweeds in an often unexpected fashion. Although there was initial resistance among some tasters to the idea of seaweed ice cream, the vast majority of them responded with satisfaction upon actually consuming the end product.

Not content to make this gastronomic innovation available to only a limited circle of tasters, Lars found an interesting way to communicate it to a wider audience. He literally took it out into the streets of Copenhagen to introduce it to the general public. During the summer months, a special ice cream bike, designed and operated by interns from the Nordic Food Lab, drove around with samples of the dulse-flavored ice cream.

▸▸ Dulse ice cream.

White onion and cod in green pepper and kelp broth   *serves 4*

Chef Josean Alija at Restaurant Nerua, located in the celebrated Guggenheim Museum Bilbao in the Basque Country of Spain, began his career at the age of seventeen. Under the tutelage of great contemporary masters in the restaurant industry, he has been able to cultivate his own style. He has developed a purist approach, in which aromas, textures, and flavors are the main components, complemented by an avant-garde approach.

Josean, true to his love of challenges, intends to dazzle those who savor his work with the essence of the earth and its produce: "In order to understand where we are, the nature that surrounds us, we must be able to understand the products nature offers; we must study today and yesterday and analyse tomorrow. By studying I realise the importance of gastronomy. I am responsible for finding and selecting products as the starting point, for innovating by tracking the unexplored; this allows me to enjoy gastronomy, not forgetting our history but continuously evolving with total freedom."

In this recipe, Josean uses the tasty kelp species oarweed (*Laminaria digitata*) to flavor a broth to go with a stewed onion covered with a softened cod skin so that it mimics the appearance and texture of a real piece of cod flesh.

- *Kelp broth:* Place 300g salted kelp (*Laminaria digitata*) in bowl with abundant water for 5 minutes to extract the salt. Discard the water. Put the kelp and 500g water together in a Gastrovac machine. Program it to run for 30 minutes after it has reached the required temperature of 60°C. Filter the resulting broth.
- *Cod and green pepper broth:* Make a green pepper and basil oil by mixing together 100g olive oil, 160g seeded green pepper, 25g green coffee beans, and 2g basil. Place them in an oven at 100°C for 1 hour. Decant to separate the oil from the liquid that has seeped out of the other ingredients. Allow the green pepper and basil oil to grow cold. Put 80g of the oil and 2g laminated garlic (cloves sliced thinly lengthwise with the germ removed) into a pot and place it near the heat. Just when it starts to boil, move it away from the heat and leave it to cool down a little. Filter. Make a cod skin broth by mixing together 400g desalted cod skins, 120g water, 25g spring onion, 3g crushed garlic gloves, and 1 cayenne pepper; keep the mixture at 62°C for 4 hours in a Roner machine. Filter. Warm up 200g

◄ White onion and cod in green pepper and kelp broth.

cod skin broth, whisk it to add air to it, and add the green pepper and basil oil very slowly. Salt to taste with fine salt.

‣ *Stewed onions*: Stew 4 onions for 1 hour in 15% brine, then drain. Put the onions into a vacuum bag together with 200g kelp broth. The weight ratio of liquid to onion is 30%. Fill the bag completely and cook in an oven at 100ºC steam on the rack for approximately 90 minutes (depending on the onions). They must be whole, with a mellow, yet slightly firm texture. Take the onions out of the bag and leave them on the rack for 10 minutes. Then place them in an oven for 30–40 minutes at 80ºC to dry slightly.

‣ *Confit cod skins:* Make cayenne pepper oil by infusing some cayenne peppers in 20g olive oil. Spread the cayenne pepper oil on 2 cod skins and place them between two layers of sulphurized paper. Put in a vacuum bag and fill it completely. Cook them in an oven at 100ºC steam on the rack for 25 minutes. Open the bag, remove the skins and cut them to the required size.

‣ *Finishing:* Peel the outer skin of the onions. Cut to resemble a piece of cod. Warm up under steam for 2 minutes. Place the onions over holm oak coals until they acquire a smoked taste. Put some cod skin on each onion piece, so that it looks like a portion of cod with its skin still on.

‣ *Plating:* Put a few spoonfuls of the cod and green pepper broth on the plate. Place the onions on top and add a few drops of cayenne pepper oil and of green pepper oil.

Vegetable 'avocado *foie*' with dulse and coriander juice   *serves 4*

In recent years chef Josean Alija at Restaurant Neruə has developed projects that run in parallel to his work in the kitchen, such as the Creativity Course or the creation of a temporary garden. Restaurant Neruə also has a research, development, and innovation kitchen directed by Chef Enaitz Landaburu, who explores, studies, and analyzes novel products and techniques.

In this recipe, Josean prepares a vegetable *foie gras* from avocado and dulse juice.

- *Acidified dulse and coriander juice:* Put 20g dehydrated dulse (*Palmaria palmata*) and 330g vegetable broth into a vacuum bag and fill it completely. Leave for 30 minutes at room temperature. Then put the combination into a Roner machine at 60°C for 45 minutes. Filter the resulting juice. Bring the juice to a boil and add 5g crushed coriander leaves. Move it away from the heat, cover with a film, and allow it to infuse for 5 minutes. Filter and add 10g lemon vinegar.

- *Avocado steamed in avocado leaf broth:* Mix 30 g dried *konbu* (*Saccharina japonica*) with 500g water in a vacuum bag. Fill completely and leave for 30 minutes at room temperature. Then, put the bag into a Roner machine at 60°C for 45 minutes. Filter the resulting broth, put it into a pot and bring to a boil. Add 10g crushed avocado leaves (without stalks), move away from the heat and cover with a film. Leave the broth to rest for 5 minutes, then filter pressing slightly. Put the avocado leaf broth into a Gastrovac machine at 100°C under pressure. Have the rack prepared inside for steaming. Meanwhile, peel 2 small avocados of about 180g each, remove the pits, and add salt and black pepper. When the machine reaches 95°C, open it and place avocado halves on the rack, close it again and vacuum cook. Allow the broth to boil for 15 minutes. The core temperature of the avocados when removed from the Gastrovac should be 70°C. Check their flavor and texture.

- *Finishing:* Heat the acidified dulse and coriander juice and add salt to taste. Cover the whole surface of the avocados with hazelnut oil. Grate a little bit of nutmeg on top. Put 3 grains of Guerande salt and 4 coriander flower petals on top of each.

- *Plating:* Carefully transfer the avocado halves to soup plates. Pour acidified dulse and coriander juice into a small jug that the server will bring to the table and pour on the side of the avocados in front of the customers.

▸ ▸ Vegetable 'avocado *foie*' with acidified dulse and coriander juice.

Sea vegetables from Maine

In 1971 Shep Erhart and his wife, Linette, began to harvest seaweeds at Frenchman's Bay in Maine, where they had settled after giving up the idea of becoming farmers. Shep told me that they had always eaten seaweeds and, at a certain point, they concluded that it should be possible to make use of the local algae instead of expensive versions imported from Asia. At the time, the macrobiotics movement was gaining converts and some of their friends were eager to obtain Maine seaweeds for their macrobiotic kitchen in Boston. This modest beginning was the basis for what would become Maine's biggest producer of seaweeds for human consumption.

As time went on, first the Erharts' kitchen, and then their barn, became too small for the thriving business. Today, they run Maine Coast Sea Vegetables, which has its own building and 20 employees who transform the locally harvested seaweeds into more than 20 different products. The raw material is delivered by about 60 seaweed harvesters who work along the coasts of Maine and Nova Scotia, where algae are found in abundance. Maine Coast Sea Vegetables processes about 50 tons of dried seaweeds annually, of which about 60% is dulse. The rest is made up of wild Atlantic *konbu* (*Saccharina longicruris*), winged kelp (*Alaria esculenta*), bladder wrack (*Fucus vesiculosus*), laver (*Porphyra* sp.), sea lettuce (*Ulva lactuca*), and carrageen (*Chondrus crispus*).

I first encountered products from Maine Coast Sea Vegetables in a health food store in Canada and, in the course of the years, I have become a great fan of their applewood smoked dulse. So it was with great expectations that I set off to visit the enterprise, which is located in the little town of Franklin on the innermost part of the Maine archipelago. The area has one of the world's most abundant occurrences of algae. They thrive in the cold, clear, and nutrient-rich waters, which are regulated by the cold Labrador Current. The tidal fluctuations along the rocky Maine shoreline are significant, ranging from three to six meters. For marine algae, it is a veritable paradise.

Shep met me and my wife late in the afternoon in August and showed us around the plant, which is housed in a large wooden structure that has a labyrinth of floors that are off-set from each other. The entire operation is marked by Shep's particular approach to seaweeds and running a business along certain ethical principles. It centers on harvesting and processing by hand, sustainability, and respect for Nature. This is also expressed in Shep's having been the first in the United States to elaborate national standards for organic harvesting and processing of seaweeds. In addition, he was a founding member of the Maine Seaweed Council, which proactively is trying to encourage Maine to regulate the seaweed harvest in order that the marine resources will be harvested sustainably.

Maine Coast Sea Vegetables is a low-tech operation, where most of the work is carried out by hand. Shep dislikes machines, so 'mechanization' is limited to some grinding mills that cut up and make granules from dried seaweeds and a pizza oven for baking crunchy energy bars made from sesame seeds, rice syrup, and kelp. People in California love these crunchy bars and Shep could sell many more of them, as well as of his other products, than he can, or wants, to produce.

Shep is very conscious of the importance of preserving the special character of Maine Coast Sea Vegetables. The business should not grow either too much or too fast. He admits that it is only pressure from his loyal seaweed harvesters that has driven an expansion of about 10% in the past couple of years. He is not particularly interested in running a large business and is not at all enthusiastic about shipping the products far from Maine. He often turns down large orders in favor of small ones and he would rather sell more locally. Shep reasonably asks why he should send seaweeds to California and Japan where there is a plentiful supply.

Shep has been in the trade for many years and trains the harvesters himself. It is of utmost importance to him that they understand the principles of collecting the different types of marine algae sustainably in such a way that they do the least harm to the environment and that no unhealthy competition springs up between them for the same pieces of coastline. In Maine the intertidal zone is open to the public and for $50 anyone can buy a state licence to harvest seaweeds. It is, therefore, vital to coordinate the harvesting of the various areas, many of which lie on the Canadian side of the border. Forty of Shep's harvesters are from the Canadian part of the Bay of Fundy, where the small island of Grand Manan has an especially good growth of dulse.

▲ Map of Frenchman's Bay, home of Maine Coast Sea Vegetables. The licence plate of Shep Erhart's delivery van.

Even though the work of gathering seaweeds is not particularly complicated, it is difficult for Shep to recruit new harvesters. This is because he requires them to dry the seaweeds themselves, a job that necessitates a well-developed ability to choose high-quality seaweeds. Drying takes place out of doors in the sun or in special drying ovens. The dried product can then easily be stored and transported to Franklin.

Shep is very preoccupied with spreading the message about seaweeds as a healthy food. He gives talks on the subject in schools and hospitals, so that everyone can learn to improve the quality of his or her diet by adding seaweeds to it. At the same time, he is only too well aware of the feelings of insecurity that are generated in the small coastal communities when there is talk of increasing the seaweed harvest. There are glaring examples of how the unregulated collection of mussels and sea urchins has altered the ecological balance of the coastal inlets. This is why Shep attends meetings with the local people to tell them about how one can ensure that the seaweed resources are harvested sustainably so that the balance is not disrupted.

Maine Coast Sea Vegetables is especially famous for its dulse. Eating it is an old tradition in Maine, brought to its shores by settlers from Wales, Ireland, and Scotland. When the dried dulse is brought to the factory, it is sorted by hand and epiphytes, shells, and small crustaceans and bivalves are picked off. The bone-dry dulse is placed in a sealed room to absorb moisture for a day and a night and then left to ripen for a couple of weeks. In the course of the ripening process, the seaweeds' own enzymes tenderize the blades so that they become both softer and more flavorful. In tightly sealed packages, these chewy, but still dry, blades have a shelf life of about a year. If they are kept longer than that, the enzymes break the dulse up into pieces and it becomes unsuitable for eating. A great deal of care is taken with the processing, which is the secret behind Shep's famous dulse, some of which is lightly smoked over applewood in a special smoke house. I like it so much that I eat it as if it were candy.

One of the most scenic wilderness areas of the United States, Acadia National Park, is located on Mount Desert Island and a few smaller islands, just south of Franklin. Set on a spectacular rocky cliff at the southern tip of the island is the picturesque and much-photographed Bass Harbor

Head Lighthouse, which has become an iconic American image. At high tide we were picked up from the nearby jetty by Erik, who runs a small motorboat service bringing passengers, mail, and supplies to Gotts Island, a tiny speck of land about 20 minutes sailing time from the mainland. Here, a new seaweed adventure awaited us.

For the past 40 years, Tina and John Gillis have spent their summer holidays on Gotts Island, which in earlier times was home to a small, hardscrabble community of fishers and lobster men. Lobsters still play a major role in the area and, on the outbound journey, Erik had to steer carefully between the hundreds of brightly colored buoys that marked the location of the traps.

Nowadays, there are only 'summer people' on Gotts Island and only a few of the hardiest among them would think of staying over the winter. All goods have to be sailed over to the island and all garbage taken away from it. There is neither electricity nor telephone service on the island. The few who come here live rather primitively in old, wooden houses. Tina and John had invited us to spend a few days with them to experience this distinctive lifestyle.

The coastline around Gotts Island varies, with low beaches facing the mainland and wild, rocky cliffs on the exposed Atlantic side. There is an abundance of seaweeds. At low tide we could wade in the knotted wrack, which forms a thick layer that, in many places, covers the purple carrageen. Wild Atlantic *konbu* (*Saccharina longicruris*) and sea tangle (*Laminaria digitata*) were also to be found. On the small land bridge that connects Little Gotts Island with Great Gotts Island at low tide, I discovered a carpet of *Porphyra*. We quickly harvested a great quantity of *Porphyra* and carrageen, which I dried on the old, sun-soaked granite step at the entrance to the Gillis family home.

The next day for breakfast I cooked *Porphyra* to make laverbread following the traditional Welsh recipe. We spread the thick paste on toasted bread. It was a delicious mouthful that was a perfect match for the fresh sea air and morning mist covering Gotts Island—it truly could be called God's island.

▲ Early morning on Gotts Island, on the rocky shoreline of Maine.

Seaweeds for industrial uses

Seaweeds as additives and stabilizers

Seaweeds for industrial uses

SEAWEEDS TURNED INTO ADDITIVES

Before going any further, it is important to set the record straight. Additives, *per se*, have a negative image and, especially in connection with foods, many consumers think of them as artificial. This despite the fact that many of them are completely natural products that have been in household use for hundreds of years without anyone giving it a second thought.

Gelling agents derived from seaweeds are a good example. When we use these substances it is simply evidence that we have extracted precisely those natural elements from the seaweeds that are important for thickening and stabilizing foodstuffs.

GELLING AND HYDROGELS

A large proportion of the global seaweed harvest is dedicated to the manufacture of gelling agents, which are used as additives and stabilizers in food and drinks, cosmetics, and pharmaceutical products. There are three classes of these substances, namely, alginate, carrageenan, and agar, all of which are polysaccharides. Under the right conditions, these substances can bind significant quantities of water in what are known as hydrogels. The hydrogels are very stable and viscous, both properties that can be exploited for technical applications.

Hydrogels derived from seaweeds are especially widely employed in the food sector, where they are used in meat and fish products, dairy products, and baked goods. Stabilizing and thickening liquid foods in the form of gels help to improve the mouthfeel of the products. Marine algae hydrogels are derived from their carbohydrate contents.

Other well-known gelation agents used in the kitchen are pectin, extracted from the cell walls of terrestrial plants, typically the peel of apples and lemons, and gelatine, derived from protein found in the bones and connective tissues of animals. Since gelatine and seaweed extracts make equally strong hydrogels, agar and carrageenan are especially useful as they are acceptable to vegetarians and vegans.

ALGINATES

Alginates are extracted from the cell walls of brown algae and are produced mainly in the United States, Norway, China, Canada, France, and Japan.

The seaweeds are treated with acid to draw out the alginates, in the form of alginic acid. This is converted to sodium alginate by the addition of soda and sodium hydroxide. As sodium alginate is water soluble, it can be separated from the insoluble fiber and proteins. The pure sodium alginate can then be reconstituted as alginic acid or calcium alginate, both of which are insoluble in water and, therefore, lend themselves to storage and transport. Sodium alginate can easily be recreated from either of these with acid treatment.

Hydrogels made from sodium alginate are used in the food industry as thickeners, gelling and binding agents, emulsifiers, and stabilizers, for example, in fish and meat products. A major end use is for fruit and dessert jellies and puddings. Specialty uses include stabilizing onion rings, sauces, pie fillings, and baked goods, as well as making margarine with a low fat content firmer.

Sodium alginate can help to ensure that foods such as pasta keep their shape when they are prepared. Its ability to absorb water and the good, mechanical stability of the gel protect the food from breaking apart or dissolving in the water. In this way, the addition of alginates can compensate for a lower gluten content in pastas.

Another targeted use of alginates is stabilizing ice cream, where it counteracts crystal formation and prevents the fats in the mixture from separating from the water. Alginates help to stabilize the foam in beer. Due to their excellent gelling properties, alginates are used in the manufacture of personal care products such as toothpaste, soaps, shampoo, and shaving cream. Recently it has been found that alginates boost battery storage by providing structural support to the electrodes without reacting chemically with the electrolytes.

Other industrial end uses are as surface coatings for the production of paper, adhesives, and textiles. Sodium alginate has now substantially replaced starch in textile printing. Because alginate thickens the dye without affecting the reaction between the dyes and the fabric, the resulting imprints are softer, the colors more vivid, and the edges sharper. In paper making, alginates help to stiffen the paper slurry and result in a paper that has a smoother and glossier surface. In addition, the resulting paper is stronger, more flexible, and has improved printability.

Gelling properties of alginates are dependent on the presence of calcium ions. One can take advantage of this relationship by mixing alginates with various fruit and berry purées. When a calcium solution is injected into the mixture, it can be made to solidify into distinct shapes, for example, as artificial cherries or long, decorative strips.

The alginate industry—
a Norwegian success story

A very extensive coastline with thousands of rocky islands, deep fjords, and clean, nutrient-rich salt water has placed Norway in the position of having the best conditions in Europe for cultivating seaweeds in the wild, sustainably, and in large quantities.

The high salt concentration in that part of the Atlantic Ocean, ca. 3.5%, leads to a reduction in ice formation in the winter and, consequently, the cold weather causes less damage to the kelp forests. The most profuse seaweed growth is found along the west coast in the middle and northernmost part of the country. It is estimated that, at any given time, there are still at least 15 million tons of wet seaweeds in the area, of which about 90% are below the intertidal zone.

Brown, green, and red macroalgae all grow along the Norwegian coast, with brown algae making up the largest biomass. At present, the brown algae tangleweed (*Laminaria hyperborea*) and knotted wrack (*Ascophyllum nodosum*) are the most important commercial species. Tangleweed alone accounts for about 80% of the total seaweed mass, which is estimated to cover more than 10,000 square kilometers of the rocky seabed on which it thrives. This works out to about 10–20 kilograms of marine algae per square meter.

In common with other European lands, the manual collection and harvesting of different species of brown and red algae in Norway were important from the Middle Ages up to about 1700. Seaweeds were used as food for humans and animals, as field fertilizer, and for the production of salt, potash, and iodine from seaweed ash. It is estimated that in the 1600's more than 100,000 tons of fresh seaweeds were collected annually and burned to produce seaweed ash.

In the 1930's, Norway again looked to this marine resource, initiating a large-scale industry to produce animal fodder from knotted wrack. At its peak in 1965, there were about 30 factories with a total output of 15,000 tons of feed supplements for cattle.

The economic impact of seaweed production took a major step forward during World War II when alginate extraction was industrialized. At first, tangle (*Laminaria digitata*) was harvested, but later tangleweed (*Laminaria hyberborea*) became the primary source. Since the start of the 1970's, a fleet of specially equipped small boats has mechanically harvested 100,000–175,000 tons of fresh seaweeds each year. Norway accounts for the largest commercial exploitation of brown algae outside of Asia and the output is exclusively derived from natural growth. Currently, there is a single company, FMC BioPolymer A/S, that is responsible for alginate production in Norway. It now accounts for two-thirds of all the alginates manufactured in Europe. In 2009, this industry contributed almost US $ 100 million to the economy.

The interesting questions then become the extent to which this encroachment on the natural growth of brown algae will disturb the ecosystem and whether the seaweed industry is sustainable in the long-term. The answer to the first is that the harvest is cautiously estimated to amount, at most, to 2% of the standing stock. This figure is significantly smaller than the total loss due to natural attrition and the harvest probably helps to renew the seaweed population and make it more uniform. So, clearly the answer to the second question is yes. Furthermore, the practice of sectioning the areas from which the seaweeds are picked, and doing so on a five-year cycle, assures that the kelp forests will have time to regenerate sufficiently.

At present, seaweed aquaculture is not practiced in Norway, but work is underway to establish it. In the meantime, the Norwegian seaweed success story shows that, with careful management, an industry based on naturally occurring seaweed populations can be run in a way that is both economically sound and sustainable.

◄ Boat specially designed for the harvesting of brown algae along the Norwegian coastline.

CARRAGEENANS

Carrageenans are derived from red algae. A significant proportion of the carrageenans on the world market are produced from the collection of wild red algae on the east coast of Canada and in Maine, Brittany, Spain, and Zanzibar. Some supplies of red algae are sourced from aquaculture carried out at the level of a cottage industry. For example, in Zanzibar many women support their families by farming algae on strings in shallow waters. In addition, the aquaculture of red algae in Indonesia is a growth sector and now exceeds its production in the Philippines.

Seaweeds for industrial uses

After the red algae are harvested, they are cleaned and dried, and the carrageenans are extracted using chemical separation techniques. The traditional, but expensive, method consists of drawing out the carrageenans using alcohol after the seaweeds have been treated with a warm, alkaline solution. A more recent process involves using potassium chloride to precipitate the carrageenans, which are then dried and granulated.

As a result of its protein-binding properties, carrageenan is used in the food sector as a stabilizer in dairy products, including cream cheeses, milk desserts, and ice cream. A mere 0.02% of carrageenan in ice cream ensures that it melts more slowly and that the milk proteins and whey do not separate from each other. In recent years, this ability to hold proteins and liquids together has been exploited for the production of 'designer fat' in meat products, for example, in hamburgers prepared from lean meat, where the carrageenan serves to retain the juice in the meat. At the same time, the carrageenan brings out the agreeable mouthfeel of an oil-water emulsion, even when the fat content is low.

Carrageenan can be incorporated into flour products and bread, where it adds body and retains moisture without having an effect on the gluten structure of the dough or inhibiting the action of the yeast.

In recent years, some researchers have pointed out that the byproducts of breakdown of carrageenans with low molecular weight may lead to conditions favorable to infections in the intestinal walls of animals and humans.

Carrageenans have a number of industrial uses. They are used to stabilize paints and cosmetics. They are also added to shampoos, where they have the advantage that they have no effect on their scent, and to toothpaste, where they cannot be broken down by the enzymes in saliva. In the same way as alginates, carrageenans are used for paper making and textile printing. On account of the fluid properties that carrageenans impart to liquids, they also

▸▸ Cultivation of Spinosum (*Euchema denticulatum*) for carrageenan production in Zanzibar.

The production of carrageenans— another seaweed success story

The small village of Lille Skensved, located about 40 kilometers south of Copenhagen, Denmark, is the home of a seaweed enterprise that has an economic impact equal to that of the *nori* fisheries half a world away in Ariake Bay, but in other ways is its polar opposite. Not only is the CP Kelco factory here the world's largest manufacturer of pectin, it is also a major producer of carrageenan, responsible for ca. 20% of the world output, which is worth about US$ 400 million annually. This operation is as hi-tech as the one in Japan is dependent on traditional, small-scale methods and its end products are intended for export, while the *nori* sheets are largely sold on the domestic market.

It is purely an accident of history that the plant is located in Denmark. All the raw materials, namely, citrus peels for pectin and marine algae for carrageenan, are imported from abroad. And less than 5% of the end product is destined for the Danish market.

The story starts in 1934 with a single individual, Karl Pedersen, who set up a marmalade factory, but quickly discovered that it was more profitable to produce pectin that he could sell to other jam factories as a gelling agent. And so, the Copenhagen Pectin Factory (CP) was born. In 1948, the company moved to Lille Skensved and carrageenan was added to the product line in 1960, during a time when the international demand for carrageenan was skyrocketing. The family-owned firm was eventually sold to American interests and since 2000 has served as the European headquarters of CP Kelco, a company with world-wide reach. It has grown from employing about 50 workers at the end of the 1940's to more than 400 today.

When I visited the factory, I met their carrageenan expert, Brian Rudolph, who has been with the firm for over two decades and also has personal knowledge of many of the places around the world where their raw materials are harvested. He told me that what is unique about CP Kelco is the ability to tailor-make carrageenans to the needs of its customers for use in food products, cosmetics, personal care products, and pharmaceuticals.

All of the carrageenans produced at CP Kelco are extracted from seaweeds that are sourced primarily from Indonesia, Chile, the Philippines,

Zanzibar, and Canada. In the Philippines and Zanzibar, two warm water species are cultivated on small family-run farms under the trade names Spinosum and Cottonii (*Euchema denticulatum* and *Kappaphycus alvarezii*). In Canada, on Prince Edward Island and in Nova Scotia, the local population harvests wild growths of carrageen (*Chondrus crispus*). CP Kelco also utilizes Gigartina (*Gigartina radula*) from southern Chile for the manufacture of specialty products for the dairy industry.

After harvest, the seaweeds are sun dried locally to remove about half of the moisture, packed into bales, and sent by container to Lille Skensved. The partially dry marine algae can be warehoused for at least a year.

At CP Kelco, carrageenans are extracted using the traditional alcohol extraction method. In principle, this process is so simple that one could do it oneself in one's kitchen. Consequently, the market is very competitive and making a profit depends on the value added that comes from intellectual property, in this case, having the expert knowledge needed to isolate carrageenans designed for specific end uses. As a result, CP Kelco has some 200–300 different carrageenans in their product line.

Carrageenan production consumes a great deal of energy to heat water and to recover the alcohol used in the extraction process. Consequently, CP Kelco has an obligation to implement environmentally friendly and sustainable manufacturing practices. To do so, it has installed its own highly efficient power plant and has invested in the largest wastewater treatment plant in Denmark in order to recycle the water. After the carrageenan has been extracted, the seaweeds are used as supplementary fertilizer on the nearby agricultural lands.

▲ Washing of Spinosum prior to the extraction of carrageenan.

◄ The red alga Spinosum (*Euchema denticulatum*). The seaweeds are harvested in Zanzibar and arrive at the factory partially dried, typically containing about 30% water.

have a promising future in ink-jet printing of textiles, which allows one to dispense with masking as part of the process.

The significant water content in carrageenan-based gels means that dissolved, smaller molecules can diffuse freely. For this reason, these gels have biotechnological applications in laboratory cultures as fixatives for enzymes and cells that need to have easy access to the substrate and nutrients.

Seaweeds for industrial uses

Agar

Agar in derived from red algae. The global demand is supplied primarily by Chile, India, Mexico, California, South Africa, and Japan. Agar is produced by cooking the seaweeds and then freeze drying the filtered, warm liquid. The finished product is sent to market as thin filaments or in granules. As it has no taste, smell, or color, agar makes a versatile food additive. In addition, it is virtually indigestible by humans and, therefore, contains no calories.

Agar can form gels by being soaked in cold water that is then heated to the boiling point. The gel sets when the liquid has been cooled to under ca. 38°C. Once set, the gel does not melt again at this point and needs to be heated to at least 85°C before melting starts. This is why foods that have been thickened with agar, as opposed to regular gelatine, have a firmer mouthfeel and retain their shape.

Agar is added to a variety of foods in the same manner as carrageenans, primarily to thicken and emulsify them. When making desserts with fruits containing certain enzymes, known as proteases, that break down proteins but are ineffective against seaweed polysaccharides, agar has the advantage over gelatine. Among others, this is true for those made with papaya and pineapple, which contain the enzymes papain and bromelain, respectively. The drawback in using agar and carrageenans is that the resulting jellies are less clear and have a coarser texture than those made with pectin and gelatine. And they do not melt in the mouth.

Another well-known use for agar is as a growing medium for microorganisms, for example, for the cultivation of bacteria in a laboratory.

Seaweeds for technical uses

Gunpowder, soda, and early glass making

The first large-scale commercial and technical exploitation of kelp occurred in California during World War I. Kelp forests were rather brutally ripped up from the seabed and the seaweeds dried and burned. A variety of salts, for example, potash (potassium carbonate), are found in the resulting ash. In combination with nitrate compounds, potash undergoes a chemical reaction to make saltpetre (potassium nitrate), which is the principal component of gunpowder.

Until the beginning of the 1800's when it was discovered how to make soda (sodium carbonate) from other readily available salts, burned seaweeds were the main raw material for soda production. Among other uses, soda is an important component in making glass, which is produced by melting quartz sand (silicon dioxide, SiO_2) together with soda and calcium, forming glass, a mixture of sodium silicate and calcium silicate. The soda also lowers the melting point of quartz, resulting in energy savings.

SEAWEEDS FOR TOOLS AND TEXTILES

Even though seaweeds are soft and quite perishable, the native people of North America collected the long, thin stipe from the brown alga *Nereocystis* to weave fishing lines. The large air bladders were used for storing water or oil.

Insoluble alginate (typically calcium alginate) seaweed fiber can also be used in the same way as those derived from plants to make textiles. They are often mixed with other fibers, for example, cellulose, silk, or polyester. Yarn produced from these seaweed blends are very soft and easy to dye.

ALGAE FOR THE PRODUCTION OF ENERGY

Dried seaweeds have been used as fuel throughout the ages even though they have low heating values and are not a particularly good heat source. As oil and energy prices have increased, however, more attention has been focused on the exploitation of seaweeds and algae for the production of biofuels such as bioethanol. Biofuels have the advantage that they can be carbon neutral and, better yet, the cultivation of algae can actually increase the rate at which carbon dioxide is removed from the environment. The waste products from biofuel manufacture can be used as animal fodder or fertilizer. The question remains about whether the production of biofuels can be made profitable, while at the same time ensuring that the carbon dioxide generated in making them does not exceed the seaweeds' capacity to fix carbon dioxide.

Regrettably, the day when algae can be put in the gas tank does not seem to be just around the corner. Either fermentation or enzymatic conversion is required to transform the algal carbohydrates into bioethanol or biodiesel.

▸ Experimental cultivation of sea lettuce (*Ulva lactuca*) for bioethanol production.

But as the carbohydrates in algae and seaweeds are far more complex than the starch in plants, the processes required to convert them into ethanol are complicated and have not been perfected on an industrial scale. In many instances, there is not yet any detailed knowledge of the biochemical principles involved in the breakdown of complex carbohydrates. Further studies are needed to elaborate effective, biotechnological production systems to bridge this gap. A number of research projects have shown that on a small scale one can use both microalgae and seaweeds to produce biofuels, but commercial viability has been achieved in only a single instance on small, family-run seaweed farms in the Philippines. In general, it has proven to be exceptionally difficult to scale up the results obtained in experimental pilot projects.

At first glance it would appear that the cultivation of microalgae would be most practical for the production of biofuels. They are probably better suited for conversion into biodiesel on account of their lipid content, which might be increased by growing genetically modified varieties. A drawback, however, is that their growth rate is often vastly overestimated. Just the same, there is optimism in some quarters, an optimism fueled by goverment grants. A small Florida company, Algenol, is developing a method for capturing ethanol directly from the microalgae without killing or fermenting them. Particular strains of algae, kept in special bioreactors containing seawater and added nutrients, are exposed to sunlight and carbon dioxide from the atmosphere or power plant emissions. The algae photosynthesize, leading to the formation of sugars inside the cells and these sugars, still in the cells, are turned into ethanol by the algae themselves. The ethanol diffuses into the growth medium, from which fuel-grade alcohol can be extracted by distillation.

Surprisingly, the most recent research has shown that macroalgae may be a more economical raw material for bioethanol than microalgae. As seaweed carbohydrates also have other industrial applications, as discussed above, it is, therefore, necessary to determine which end use is most economical and whether several of them could be combined.

Brown macroalgae contain the sugar alcohol mannitol and the glucose polymer laminarin, which can be fermented to alcohol, but only about 25% of their carbohydrates can be converted to ethanol. In particular, their alginic acid content cannot be converted to ethanol, but only to methane using an anaerobic process. Attention has, therefore, turned to the green algae, which contain few gelling agents that are of interest but instead have large quantities of starch-like carbohydrates that can be fermented directly into alcohol. Among them, sea lettuce, which grows rapidly, yields a greater volume of ethanol, especially when it is harvested young. Sea lettuce can be cultivated on land in tanks, possibly making use of nutrients that are waste products from agriculture or aquaculture and heat from water that is being recycled for cooling from power plants. Such tank farms would, however, take up vast surface areas in order to receive sufficient light for growth.

In order to establish a large-scale biofuel industry based on algae, then, one has to calculate the significant outlays required for construction of the actual facility, for provision for a nutrient supply and temperature control for algae cultivation, and for the operation of extraction or distillation processes. Moreover, the energy required to build and run such a plant results in carbon dioxide emissions. This means that, even if oil prices rise, there has until now been a substantial gap between the known costs of producing energy from conventional oil and gas sources and the projected costs associated with biofuels from marine algae. There is nothing to indicate that this differential will be eliminated anytime soon. And should such biofuels become commercially sustainable in the near future, there is, at this time, no reason to suppose that they would account for a large proportion of our energy needs.

On the more futuristic side, Japanese researchers and private companies are planning an enormous project to establish seaweed aquaculture over a huge area of 10,000 square kilometers in the middle of the Sea of Japan. The proposal envisages a yearly output of 20 billion liters of ethanol, equivalent to a third of Japan's annual gasoline consumption.

FOOD OR FUEL?

The global demand for energy is unlikely to decline, regardless of the degree of energy efficiency that it is possible to achieve in the developed countries. Robust, new economies in Asia are expanding by leaps and bounds and the Third World has a large, unmet demand for energy sources. At the same time, a hungry world needs more food. It is, therefore, quite possible that our deliberations about the industrial utilization of algae will be confronted by the same dilemma as the one that has arisen with respect to the growing exploitation of plants, such as corn, soya, and rapeseed, which are often the dietary staples of poor people, for the extraction of ethanol.

There are indications that rising prices and the looming global food crisis are partly due to the diversion of agricultural output from foods to energy production. This raises the fundamental question of the extent to which we should aim to dedicate marine resources to increasing the food supply or to trying to fill the world's insatiable demand for more energy. Another basic question is whether the exploitation of seaweeds and algae for energy production can be carried out in a sustainable manner, so that their capacity to fix carbon dioxide, with its inherent positive impact on the Earth's atmosphere and climate, is not diminished.

Seaweeds in medicine, health care, and cosmetics

SEAWEEDS IN MEDICINE

A variety of seaweed species have been incorporated into traditional Chinese and Japanese herbal medical practice in the treatment of tuberculosis, rheumatism, colds, influenza, wounds, worm infestations, and cancer. Just as it is difficult to dismiss several thousand years of experience with naturopathic medicine, it is equally difficult to establish a firm scientific basis for its beneficial effects, to say nothing of trying to shed some light on precisely which substances in the seaweeds are the biologically active ingredients and what determines their bioavailability. Over and above this, advocates of macrobiotic approaches to wellness will probably maintain that the determining factors are not the individual chemical substances, but the seaweed in its entirety and the synergy between it and the patient in a particular treatment situation.

Nevertheless, many *bona fide* researchers have come to the conclusion that various seaweed species may well contain substances that are biologically active in specific ways. Work is on-going in laboratories all over the world to try to isolate many different substances from a whole range of marine algae. In particular, the search is targeting those substances that might interact with cancers, the immune system, inflammations, viruses, and pathogenic fungi and bacteria. Until World War II and the introduction of antibiotics, iodine extracted from seaweeds was an important antiseptic and disinfectant.

The medical and pharmaceutical industries have found uses for sodium alginate extracted from seaweeds. One such application is for making dental molds. The alginate is mixed with water and a coagulant that contains calcium ions, allowing the dentist to make a neat and precise impression of the teeth.

Creams and pastes that emulsify active curative ingredients, for example, for the treatment of skin diseases, take advantage of the gelling properties of alginates. The hard capsules for medicines that are taken orally and that need to pass through the digestive system more slowly to allow their active ingredients to be released over a longer period of time often contain alginates. Carrageenan is utilized in the manufacture of suppositories and soft capsules.

Calcium alginate, which is insoluble in water, is used for making plasters and bandages to cover wounds. The alginate can be manufactured in the form of thin threads from which it is possible to make a gauze-like material. When the bandage is placed on a large, hard-to-treat wound, for example, one resulting from burns, it molds itself around it. The advantage this alginate bandage has over one made from cotton or other textile fibers is that it is easier to remove. One simply soaks it in a saline solution which converts the water insoluble calcium alginate to a water soluble sodium alginate, which can more easily be removed from the wound without damaging it.

There are several biomedicinal and pharmaceutical applications of mannitol derived from seaweeds. It contains only half as many calories as ordinary sugar and tastes less sweet. As it is indigestible and odorless, it has no effect on the glucose and insulin levels in the bloodstream and is a good, neutral sweetener. It also acts as a stabilizer in toothpaste. When used for sugarcoating tablets and chewing gum, mannitol has the advantage that it does not absorb water and, therefore, the tablets do not stick to each other. As mannitol is an osmotic diuretic, it is can also be injected intravenously to help reduce instances of elevated pressure within the eyes and the cranium.

Interestingly, it has recently been found that certain bacteria (*Bacillus licheniformis*) that live on the surface of raw seeweeds secrete enzymes that help to clear dental plaque when applied as an ingredient in toothpaste.

BIOLOGICALLY ACTIVE INGREDIENTS

Seaweeds for
industrial uses

Seaweeds and algae are ancient forms of life, which have existed unchanged for hundreds of millions of years. One might, therefore, expect that they would have evolved some unique and effective defense mechanisms. This does seem to be the case. Their defense mechanisms are founded on the presence of chemical substances that protect the algae against attack by microorganisms and herbivores. Consequently, seaweeds and algae have been singled out as a potential gold mine for the extraction of interesting bioactive substances with many possible applications in the service of mankind. Iodine is the most obvious and well-documented example.

Research has led to the identification of a whole range of chemical substances, especially the so-called secondary metabolites, which have interesting biological effects that can be leveraged for pharmaceutical and medicinal purposes. Since the 1970's, the search for these substances has intensified, in particular for ones that could have anti-cancer and antibiotic applications.

Some of the newly discovered substances are particularly intriguing. The polypeptide (kahalalide F), which in some seaweed species acts as a chemical defense against grazing herbivorous fish, has an anti-viral effect. This is a possible partial explanation for the lower incidence of HIV/AIDS in areas where seaweed consumption is common. For example, the incidence of HIV/AIDS in Japan and Korea is 0.01%, whereas in Africa it is 10%. Differences in sexual behavioral practices may account for part of this discrepancy. On the other hand, it is interesting that in Chad, where a large proportion of the population eats blue-green microalgae (Spirulina), the incidence of HIV/AIDS is only 2–4%. Pilot studies have indicated that the onset of AIDS can be delayed in those patients who have not yet been treated with conventional anti-viral medicine by supplementing their diet with seaweeds.

Carrageenan from red algae has a well-documented anti-viral effect, for example, against herpes, the HIV-virus, and the human papilloma virus that causes cervical cancer. The anti-viral effect seems to have two components. One is to prevent the virus from invading the cells and the other is to strengthen the innate immune system and impede further infection via the bloodstream. By eating seaweeds and microalgae regularly, one may attain both a certain level of protection against the virus and, for those who are already infected, a substantial reduction in the output of the virus, thus lowering the risk of contagion.

The polysaccharide fucoidan from brown algae has been shown to be able to counteract the formation of gastric ulcers by suppressing the ability

◄ Sea urchins eating seaweeds.

of its precursor bacteria, *Helicobacter pylori*, to colonize the stomach lining. In addition, brown seaweeds contain a long list of terpenes, an umbrella term for a large and varied class of organochemicals. Some of these terpenes exhibit anti-viral and anti-cancer properties and possibly have the potential to counteract malaria. As described earlier, some of the fats (certain glycolipids) and some of the polysaccharides (especially fucoidan) that have already been discovered have the ability to suppress the growth of tumors.

As the search for new antibiotics has intensified in the last few years due to the rapid proliferation of drug-resistant bacteria, attention is increasingly being paid to the substances found in seaweeds and algae that are natural defenses against attacking microorganisms and herbivorous marine fish. The brown algae contain specific tannins and the green and red algae have, for example, acrylic acid and bromophenols. These compounds exhibit anti-bacterial activity and, furthermore, discourage herbivores from eating the algae. It has been proposed that using a combination of extracts of such bioactive substances from a variety of seaweed species in connection with the aquaculture of fish, would make it possible to dispense with conventional antibiotics in the fish fodder.

A number of reports in the scientific literature suggest that extracts derived from various seaweeds can counteract fungi and mosquito larvae and, therefore, may have potential as insecticides. Currently, research is being undertaken into the use of special algae, for example, transgenic algae, which produce different substances that can serve as a type of vaccine against skin infections and sea lice on farmed salmon.

Seaweeds, and especially extracts from red algae, have a long-standing place in folk medicine as a remedy for intestinal worms in countries such as Norway, Turkey, Greece, Japan, China, and Indonesia. This effect is probably due to certain amino acids (domoic acid, kainic acid) found in a few seaweeds. These amino acids are toxins, which in larger quantities affect the nervous system, lead to paralysis, and, in serious cases, cause brain damage and memory loss. Small amounts are, however, well suited for killing the parasites.

Seaweeds in health care and cosmetics

As a result of their excellent gelling properties and the ability to bind water molecules, both alginates and carrageenans are utilized as thickeners and stabilizers in toothpaste, shampoo, and creams. An additional advantage is that carrageenans are not broken down enzymatically. Carrageenans are instrumental in maintaining an appropriate degree of viscosity and fluidity in tooth paste and also keep its taste substances and polishing agents stably emulsified. Apart from contributing stability and texture to cosmetic products for hair and skin care, carrageenans allow for the emulsification of sugar alcohols and a variety of substances that are added to the products. Carrageenans are now used in contraceptive gels, where they also appear to be able to bind the HIV-virus and thereby reduce the risk of transmitting the disease.

Seaweeds are used extensively in beauty treatments, going back at least as far as the time of the Babylonians, who found them beneficial for skin care. A special type of spa treatment called thalassotherapy (from the Greek word *thalassa* meaning sea), based on seawater and algae, is available in many places around the world. The seaweed treatment is intended to leave the skin moist, soft, and supple, an effect that is well known to those who harvest and process seaweeds by hand.

Algotherapy, that is, treatments and baths using seaweeds and their derivatives, is now available world-wide in spas and beauty clinics and is a rapidly growing sector. It is difficult to gauge the effectiveness of these treatments and to determine which algal substances may contribute to positive outcomes.

It is claimed that seaweed baths are effective against a range of ills, such as rheumatism, eczema, asthma, and stress related sicknesses, and promote weight loss to boot. Whether or not there is any scientific basis for this assertion, it is easy to ascertain that seaweed baths are associated with a bodily sense of well-being.

Seaweeds that grow in shallow water have evolved a number of pigments, which are chemical substances that, among their other functions, protect the seaweeds from the ultraviolet rays of the sun, which can penetrate the top layer of the water. Some of these pigments have been isolated and, as it was found that they absorb light in the ultraviolet part of the spectrum, they are now incorporated into sunscreen products.

◄ Sheep eating seaweeds on the Tasiluk coast in Greenland.

Seaweeds as animal fodder and fertilizer

SEAWEEDS FOR DOMESTIC ANIMALS

Seaweeds are considered to have a nutritional value equivalent to that of wheat and it is likely that one of the earliest ways in which they were used was as fodder for domestic animals. Written sources from the 1st Century CE describe the feeding of cattle with seaweeds in the Mediterranean basin. Along the coastlines of Iceland and the Faroe Islands, as well as in Norway and Ireland, different species of marine algae were used throughout the Middle Ages when there was not enough grass for the domestic animals.

The importance of seaweeds as animal fodder is mirrored in the local names of some species. For example, knotted wrack is called 'pig food' in Norway and dulse is called 'cow weed' in England and 'horse seaweed' in Norway.

On the Orkneys, the North Ronaldsay native breed of sheep still eat seaweeds at the foreshore. A wall known as the sheep-dyke was built to confine the sheep to the beach for most of the year, as the grass is reserved for raising cattle. The sheep are allowed onto the pastures for only a short time during the lambing season. During the rest of the year, they depend completely on seaweeds for sustenance. The mutton from these animals is highly regarded for its special salty flavor and sold as a gourmet specialty.

Similarly, seaweeds are used to supplement the winter fodder in Greenland and Iceland, and the sheep themselves regularly seek them out on the beach. The Icelandic firm Thorverk manufactures organic seaweed meal for animal fodder from dried sea tangle and knotted wrack.

Seaweeds for industrial uses

Inspired by the success of the Orkney sheep rearing practices, Australian researchers are undertaking studies on the effect of adding seaweed fibers, which are more digestible than those of grasses, to the cattle diet. Doing so might help to reduce the amount of methane gas generated by bovine digestive tracts. As cattle herds are estimated to account for up to 20% of the world's methane emissions due to man-made activities, and as methane is considered more damaging to the ozone layer than carbon dioxide, this could help to reduce the greenhouse gas problem and mitigate global warming.

Dried brown algae, especially knotted wrack, are used to manufacture food supplements in the form of granules or powders for domestic animals such as pigs, cows, horses, sheep, and chickens. It is usually given in portions that make up about 3% of the total fodder intake. The supplements given to milk cows can lead to an increase of 6% in the butter fat content of the milk. A decrease in the rate of infections has been observed in cows raised for meat, which may possibly be due to an antibiotic effect of certain substances present in seaweeds.

Marine algae are a component in the fodder for fish farms. In Ireland a company called Ocean Harvest Technology has recently developed a new type of natural and sustainable ingredient in salmon feed. It contains a complex blend of bioactive ingredients derived from a variety of seaweeds sourced from around the world. Farmed salmon fed on this functional food product are found to have greater weight gain, higher levels of omega-3 fatty acids, better health, fewer infections and sea lice, and lower mortality rates. As a bonus, their meat has a better texture, flavor, and appearance.

SEAWEEDS AS FERTILIZER

Seaweeds and seaweed compost have been used as fertilizer in coastal areas for hundreds of years. In France and on Iceland, this practice goes back at least as far as the 14th Century. Originally the seaweeds were collected on the beach and spread out on the fields. During the time when cereals were still grown on Iceland, the fields were often fertilized with seaweeds and algae were also spread on vegetable patches well into the 1900's.

▸▸ Knotted wrack (*Ascophyllum nodosum*) is harvested by hand at low tide on the west coast of Ireland.

Pieces of seaweed that wash up on the shore have for centuries, especially in Scotland and Ireland, been combined with good soil to create raised beds

▸ Sandy beach on the west
coast of Ireland. The reddish
color visible in the shallow
water is due to deposits of maërl
from calcareous red algae.

in which to cultivate potatoes and other crops. These so-called 'lazy beds' retain moisture quite well, but the salt and possible pollutants introduced by the seaweeds pose some challenges of their own.

Deposits of calcareous red algae, known as maërl, have been used, and are still in use to a certain extent, to condition soils that are lacking in calcium. As maërl accumulates at a painfully slow rate, the exploitation of this resource is by no means sustainable.

In recent years, those wishing to follow organic gardening practices have been advised to use seaweeds to fertilize and mulch the beds of certain vegetables, such as asparagus. Asparagus, like cabbage and celery, which were also originally shoreline plants, is salt tolerant and the seaweed mulch helps to control weeds. In order to avoid making the soil saline, which would disturb the habitat of the earthworms, the seaweeds are first washed in rain water. Also, these seaweeds must be collected in uncontaminated waters, that is, far from industrial and urban areas and storm drains.

"In the Bishopriche of Durham
the housbandmen of the
countie that dwel by the sea
syde use to fate [fatten, i.e.,
manure] their lande with
seawrake." (William Turner,
A New Herball, 1568).

Nowadays dried granules and extracts in liquid form, especially from brown algae, are used as plant fertilizers and growth stimulants. There are indications that these types of environmentally friendly agri-chemical products will take on greater importance in the face of steadily increasing opposition to the use of artificial fertilizers. Apart from their mineral content, using seaweeds as soil conditioners has the advantage that they tend to cause the dirt to clump together, which makes it more porous. Greater porosity allows more air and water to penetrate the soil, which, in turn, stimulates desirable earthworm activity. Claims have also been made, although without a great deal of supporting evidence, that seaweed fertilizers boost the plants' resistance to frost and to attack by insects and fungi.

"The seaweeds have to be there, if the children return home"

Today, fewer than 30,000 Irish men and women speak Irish fluently as their principal language. They live mainly along the west coast of Ireland in small rural districts that are called *Gaeltacht* (Gaelic-speaking). The Irish language together with Scottish (Erse) and the now virtually extinct Manx make up the Gaelic branch of the Celtic language group.

As Irish disappears little by little as a living language, a whole range of words and expressions for things and relationships that have been an integral part of the Irish way of life and that reflect the habits and activities of an earlier culture are disappearing along with them. A good example of this loss involves words for seaweeds, the harvesting of seaweeds, and foods made from them, all of which played a major role in the traditional impoverished Ireland of the past. The Irish author Aidan Carl Matthews (1956-) mourns the death of the Irish language as follows: "The tide gone out for good, thirty-one words for seaweed whiten on the foreshore."

Connemara is a *Gaeltacht* in western Ireland, north of Galway; here the common language of daily life is Irish. When I had the chance to visit Connemara, I had, however, to rely on those who spoke English. Nevertheless, Dara Flagherty used the Gaelic word *cleimín* when he pointed to a large bundle of about 2–3 tons of wet seaweeds, which had been pulled up from the edge of the water in a small bay south of Cill Chiaráin. For the past 15 years, Dara has worked as the seaweed manager of a small seaweed factory, Arramara Teoranta, which was established there in 1947. The most important part of his job is keeping on-going contact with the ca. 320 seaweed harvesters who live along the hundreds of kilometers of coastline. So when Dara said *cleimín*, he meant bundle, but not just any bundle. The word is used only in relation to seaweeds and is one of the 31 Gaelic words related to algae that are on their way to extinction.

The seaweeds that I saw lying at the edge of the water were knotted wrack (*Ascophyllum nodosum*), held together with a long rope. Dara has the challenging task of coordinating the work of the individual harvesters so that deliveries to the factory are steady, regardless of the time of year and the very changeable and harsh weather conditions on the exposed

Atlantic coast. The seaweeds must not be more than a couple of days old before they go into the dryers in the factory.

Harvesting seaweeds is a long-standing tradition in Ireland. The farmers, sheep raisers, and fishers who have land adjoining the coastline have for hundreds of years had the right to harvest the seaweeds of the foreshore and the intertidal zone to feed themselves and their domestic animals and to fertilize their plots of land and potato beds. Now they are taking advantage of their established rights to collect seaweeds for industrial ends. They harvest the seaweeds by hand at low tide, go out by boat at high tide to gather them together, bundle them into *cleimíns* with a rope, and tow the bundles to the shore with the boat. From that point on, it becomes Dara's job to organize transport along the highway to the plant. In earlier times, the bundles were moved in special seaweed boats, but now it is done in large trucks, which can each carry a load of 20 tons of wet seaweeds.

For about a hundred of the harvesters, their sole source of income is the payment they receive for the 20 tons of seaweeds they can manage to harvest every two weeks. The majority of the others have to supplement their seaweed earnings by farming or fishing. Even though the Gulf Stream tempers the severity of the winter weather, storms and rain are normal and harvesting seaweeds is hard work. Everything is done by hand along an unmarked coastline, where the tidal range can be as much as five meters. The harvester stoops to cut the seaweed free with a short knife or a scythe. This must be done carefully so that the organism can grow again for re-harvest in four to five years' time. At present the going rate is about €38 per ton of wet knotted wrack. This corresponds to ca. US $ 1,100 for the 20 tons of seaweeds that a harvester is able to collect in a two-week period.

It seems obvious that the whole operation could be made more efficient by forming cooperatives and adopting mechanized ways of harvesting or by selling the rights to a company that could streamline the harvest, increase the yield, and thereby make it more profitable. Donal Hickey, the director of the seaweed factory, smiled indulgently at these suggestions when I visited him together with the Danish seaweed researcher Susse Wegeberg. He told us that the local seaweed harvesters would never dream of making changes and would under no circumstances sell their harvesting rights. As he said, "The seaweeds have to be there, if the children return home."

His statement resonated with something that is deeply rooted in the Irish identity. Until recently, Ireland was a very poor country. From 1848 to 1950, more than 6 million Irish men and women emigrated and the population fell from 4.4 million in 1861 to 2.8 million in 1961. About one million died during the Great Famine caused by potato blight at the end of the 1840's. The tide did not turn until after the economic miracle of the 1990's, when Ireland in 1996 finally experienced a net influx of immigrants, the last European country to reach this milestone. The enormous emigrations left their mark on the Irish psyche. Entire families were uprooted and others were split apart. Those who remained clung to the hope that family members who had gone away would return again. And when they did, there should at least be something to live on—seaweeds. The seaweeds that could be fed to domestic animals and fertilize the potato bed, the seaweeds that would be the only way of putting food on the table in an impoverished household if all else failed.

The people of Ireland have always seen seaweeds as a means of survival and they are, therefore, not something that one gives away lightly or treats without respect and care. Donal Hickey explained that as long as one had seaweeds one could always get by and would not need to 'hit the road'.

Seaweeds are still the last remaining hope for the local area and the Irish government views the state-owned seaweed factory at Connemara as an important means of supporting the development of this *Gaeltacht*. Without the plant, people would have to live on welfare and then even more of the young people would leave for the large cities. Here, seaweeds are simply the only way to make a living.

The seaweed factory handles 20,000 tons of knotted wrack each year and the director estimates that the coastline could sustainably yield three times as much. Previously, the seaweeds were dried and sent, unprocessed, to Scotland and the United States, where a large international company extracted the valuable alginates from the raw material.

In order to derive a greater local economic benefit from the seaweed harvest, the factory in Connemara has in the last few years taken over processing the seaweeds to make feed supplements for animals and fish and soil conditioners. These products are now exported to twelve different countries. Donal Hickey views this low-tech business as the embryonic

means of moving toward other and more lucrative ways to use the seaweeds. "We need to know more and to change the seaweed harvesters'
outlook that seaweeds are merely something brown."

The next step should be the harvesting and, possibly, cultivation of
seaweeds other than knotted wrack, preferably species that are well suited
for human consumption. Here Ireland has the advantage that the ocean
waters along its west coast are among the cleanest in the world and it
would, therefore, be relatively easy to obtain organic certification. The
American company Sea Vegg is already taking advantage of this pristine environment to manufacture an expensive food supplement from seaweeds
imported from Ireland. Sea Vegg sells the powdered, dried seaweeds in
small capsules, selling for about US$ 30 for a one month supply.

A little to the north of Connemara in Westport, Seamus Moran, has
done something about these possibilities. In 2005 he gave up his twenty-
year career as a cook and incorporated a small family-run business, LoTide
Fine Foods. The company is based on the seaweeds harvested by Seamus
and a half dozen other local people. In the nearby bay, they collect dulse,
carrageen, winged kelp, thongweed, and various types of sea tangle, all of
which are high quality and eminently edible seaweeds.

Seamus is passionate about marine algae. Using his childhood experience of seaweeds cooked in milk for breakfast and dried algae for snacks
during the day as a starting point, he is determined to find new and contemporary ways of eating them, focusing first on taste and then on their
nutritional value. As he said, "I cannot get people to eat seaweeds just
because they are good for them." Seaweeds have to be added to ordinary
food in small quantities to enhance their taste in a carefully worked out
combination of different varieties. His solution is to make dried seaweed
mixtures, which have been so well received that they are now used in
leading gastronomic establishments.

It is a huge leap from knotted wrack spread on the potato beds of dirt-
poor coastal farmers and a breakfast of carrageen and milk to premium
quality seaweed products served in the leading restaurants of Dublin. In
western Ireland, it appears that one can make that leap without abandoning
one's roots. Here Irish is still the main language when it comes to something as important as seaweeds.

Epilogue: Seaweeds—edible, available, and sustainable for the future

In this volume, I have attempted to gather together, for the non-specialist, much of the existing knowledge about how humans have used, and continue to use, seaweeds as food, for medicinal purposes, and in a variety of practical and industrial applications.

On one level, the book is intended to show how one can, with little effort, incorporate seaweeds into ordinary dishes to create tastier, healthier, and more interesting meals. On another level, however, this volume is a reflection of my conviction that marine algae should be regarded from a broader perspective as a vital resource for the immediate future. As indicated by its title, I have paid special attention to the three characteristics that strongly support the idea of utilizing them more extensively and in a much more informed way.

Seaweeds are not only *edible*, they are full of flavor and packed with important vitamins and minerals. They make up a vast, diverse group of organisms, which are distinctly different from terrestrial plants. There are presently more than 10,000 known species of marine algae and, as they are widely dispersed all around the globe, they are readily *available* in virtually all climatic zones. In many cases, seaweeds can be exploited in a *sustainable* way, either by harvesting the algae in the wild or by farming them in very large quantities in the ocean, preferably in a sustainable multi-trophic aquaculture.

The implications of edibility, availability, and sustainability are wide-ranging. Seaweeds are a significant nutritional resource and promoting them for human consumption, in both the developed and the developing world, may be part of a solution to the world's urgent need for increased food production and, at the same time, this could result in a healthier diet and help to counteract some of the diseases related to poor eating habits. It is a historical fact that extreme conditions, such as shortage and wars, have created new opportunities for the exploitation of seaweeds.

Seaweeds may constitute an as yet barely tapped source of chemical substances for the development of new active compounds for medicinal purposes and for industrial applications. Seaweeds are also on the radar in the energy sector, where they may be a means of meeting growing global demand and of decreasing the carbon footprint that is attributable to human activities. With a view to solving some of the major problems confronting the transportation sector by finding a replacement for oil and gasoline, plans have already been drawn up in Japan to grow brown algae on a massive scale to produce bioethanol.

It is supposed that an additional benefit of this enterprise would be that the enormous seaweed forests could help to clean up the ocean by capturing some of the fertilizer run-off from agricultural and urban areas. Elsewhere, leading companies in the aviation industry are currently experimenting with how to produce biomass from microalgae for third-generation alternative biofuels to power passenger jet planes. It is noteworthy that, unlike many other biofuels, those based on algae would not displace food crops or provide competition for fresh water resources.

The economic value of the oceans' ecosystems has been estimated to be US$ 25,000 billion, and the marine algae and sea grasses to have a value that is greater than that of the rainforests and coral reefs together. Some of these economic benefits are in jeopardy and the ecosystems of our oceans are under seige. Overfishing, pollution, excess emission of nutrients, and global warming have an impact on algal diversity and the balance between different species. The 2011 Fukushima accident in Japan is an example of how radio-active pollution can affect the seaweed industry. Conservation of our precious marine ecosystems requires international collaboration and attention.

In 2010, I organized the first international and interdisciplinary conference with a specific focus on seaweeds for human consumption, manufacturing of bioactive compounds, and combating diseases. The conference attracted an eclectic mix of leading scholars, outstanding exponents of sustainable seaweed harvesting and farming, experts in the use of seaweeds for human nutrition and health, innovative chefs, industrialists, and entrepreneurs. Topics covered ranged from the scientific aspects of marine algae to practical matters such as food safety, health claims and legislation governing these, and epidemiological studies of their dietary value. While seaweeds are still a staple food in some parts of the world, are being marketed more widely, and are gaining much greater acceptance on gastronomic menus, there is still much to be done. The main conclusion of this conference was crystal clear: there is a pressing need for basic research in order for mankind to extract maximum benefits from all aspects of marine algae.

But even before results from this research become available, there is something we can all do: start to introduce seaweeds into our home cooking on a daily basis. I hope that this volume may serve as an invitation to have fun in the seaweed kitchen and be a source of inspiration and creative ideas.

Epilogue: Seaweeds—edible, available, and sustainable for the future

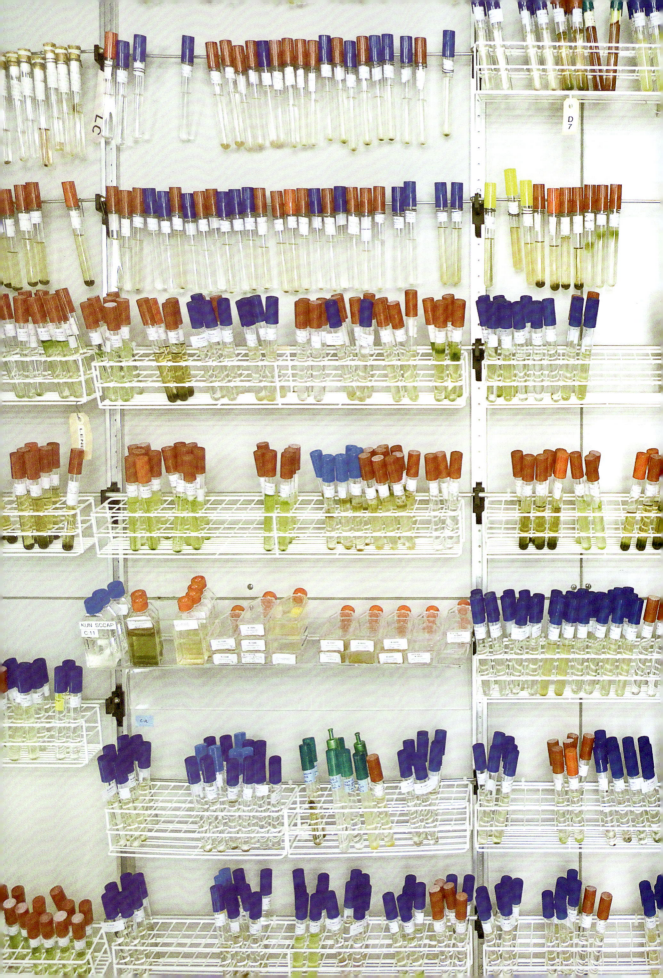

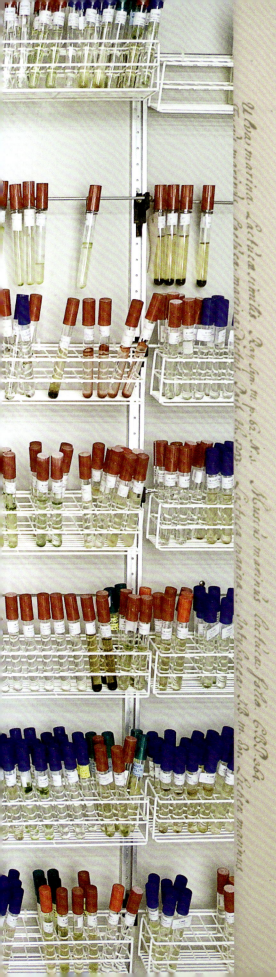

Technical and scientific details

Evolutionary history and the life cycle of seaweeds

EVOLUTION

Technical and scientific details

As discussed near the beginning of this book, the impact of algae on the global ecosystem is enormous. It is estimated that they are currently responsible for about 90% of the oxygen that is released into the atmosphere. Furthermore, their contribution to the physical conditions on Earth were vitally important in setting the stage for the evolution of higher organisms.

The first signs of life on our planet date back to a time when it was still very young. Earth was formed about 4.5 billion years ago and it is thought that the earliest organisms had already appeared more than 3.8 billion years ago. During this period, the conditions on Earth were very different from those of today. A particular indication of the physical state of the planet was the nearly total absence of oxygen in the atmosphere, less than one part in ten billion. Life consisted of simple, unicellular organisms, the so-called prokaryotes, which most closely resemble present-day bacteria. The prokaryotes encompass two separate domains (or superkingdoms): the Bacteria and the Archaea.

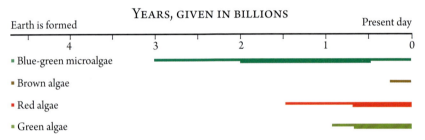

▶ The evolution of microalgae and macroalgae on Earth. The thick lines indicate times during which there was a rapid increase in the occurrence of these species. Macroalgae became prevalent about 500 to 800 million years ago.

About 2.5 to 1.5 billion years ago, there was a noticeable change in the Earth's atmosphere, as the amount of oxygen in it started to increase. This was brought about by the appearance of new forms of life that could use sunlight to convert carbon dioxide into oxygen and carbohydrates, a process we know as photosynthesis. There is probably no other development in the whole of the Earth's existence that has altered its surface and climatic conditions as dramatically as did photosynthesis. The oxygen-producing blue-green microalgae, which go back about 3 billion years, are considered to have been instrumental in effecting this change.

Chemically speaking, oxygen is a very reactive substance. Those organisms that could not adapt to its presence either died, buried themselves in

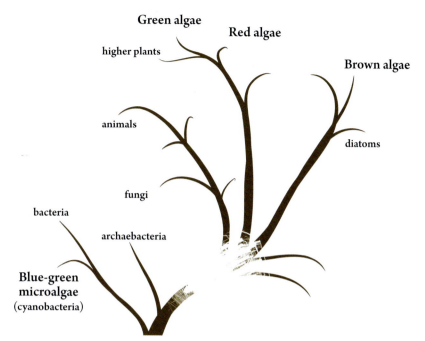

Green algae

higher plants

Red algae

Brown algae

animals

diatoms

fungi

bacteria

archaebacteria

Blue-green
microalgae
(cyanobacteria)

◄ The 'tree of life' (phylogenetic
tree). The tree in this illustration
is incomplete, but shows the
relative placement of macroalgae
and a single group of microalgae. The undefined area in the
middle marks the transition
between prokaryotes (Bacteria,
Archaea) and eukaryotes.

the seabed, or evolved into new organisms that could tolerate it. On the other
hand, the altered conditions provided an opportunity for new and more
complicated forms of life to evolve. This was the time when a whole new
domain arose, as the so-called eukaryotes appeared on the scene. Eukaryotes,
in the classical sense, encompass fungi, plants, and animals. Their survival
is dependent on respiration, that is, the use of oxygen to create the energy
needed for their life processes.

Among these eukaryotic life forms were the multicellular macroalgae, the
different genera of seaweeds. They started to appear about 1.5 billion years
ago, becoming more widespread about 500 to 800 million years ago, and
added their own contribution to the oxygenation of the Earth's atmosphere.

There are many characteristics that differentiate the eukaryotes from the
prokaryotes. One of these, namely, the presence of higher sterols in the cell
membranes of the organisms (cholesterol in the case of animals, ergosterol in
fungi, and phytosterols in plants and algae), is of special interest as it presup-
poses the existence of molecular oxygen in the atmosphere. Higher sterols
can only be synthesized by living organisms in the presence of molecular
oxygen. In this way, oxygen unleashes new evolutionary driving forces that
result in greater biodiversity.

The relationship between the various life forms is normally illustrated as
a 'tree of life', known formally as the phylogenetic tree. It is not certain that all

eukaryotes have a common ancestry among the prokaryotes and it is possible that the different main branches of the eukaryotes have arisen at different points in time. Despite advances in our knowledge, there are many areas of uncertainty with regard to the details of the 'tree of life' and, consequently, many gaps in our understanding of the evolution of the various types of organisms. As more precise information about the genetic make-up of the different species becomes available, the evolutionary path is clarified and this leads to changes to the 'tree of life' and to the classification of organisms.

From the illustration, it is clear that the green and red algae are more closely related to each other than either is to the brown algae. In addition, the green algae are closely related to the higher plants, for example, flowering terrestrial plants. At present, consideration is being given to abandoning the old classification for the plant kingdom, Plantae, where red and green algae belong. Instead, this kingdom is to be designated as the archaeplastids, Archaeplastida, that is to say, organisms with primary chloroplasts. They are distinguishable from each other by the type of chlorophyll that they contain. Under this scheme, brown algae, together with various groups of microalgae (including the diatoms), are included in another kingdom and are now classified as heterokonts, which are characterized by some special properties of their male reproductive cells.

THE LIFE CYCLE OF SEAWEEDS

Several distinctive types of life cycles, some of them very complicated, are associated with the different varieties of seaweeds. Typically, they involve alternation of generations (also called alternation of phases), which have either a single set of chromosomes (haploid, gametophyte generation) or two sets of chromosomes (diploid, sporophyte generation).

With some types of algae, for example, *Ulva*, the alternation of generations is isomorphic, that is to say, there is no change in their outward appearance. With others, like *Porphyra* and *Laminaria*, the alternation of generations is heteromorphic, meaning that the different generations do not resemble each other. Among the life cycles of the heteromorphs, there are some types, such as *Laminaria*, where the differences are striking. Its sporophyte generation is the well-known organism with large blades, while its gametophyte generation consists of a microscopic, filamentous body. For other types, including *Porphyra*, the reverse is true and the seaweed with the large blades is the gametophyte generation.

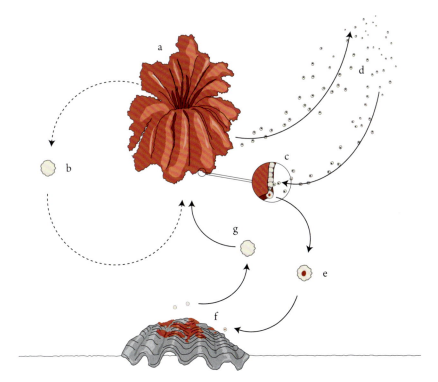

◀ Life cycle of the red alga, *Porphyra*.

THE LIFE CYCLE OF *PORPHYRA*

The red alga *Porphyra* has an especially complicated life cycle, with a fascinating aspect that merits further discussion because of the interesting history associated with its discovery. The latter was the essential piece of information that facilitated the large-scale commercial exploitation of this particular seaweed for the production of *nori*, an industry that has an enormous nutritional and economic impact and that is one of the most successful examples of aquaculture.

Nori is made from the macroscopic large blades of the haploid generation of *Porphyra* (a in the diagram above), which can reproduce itself by shedding asexual spores (b). The blades can also produce egg cells (c) and sperm cells (d). The egg cells remain on the blade where they can be fertilized by the sperm cells, resulting in a zygote (e) that is released. This diploid zygote forms what are known as carpospores, which germinate to produce a microscopic, filamentous and branched generation. By way of an enzymatic process, these microscopic organisms can bore themselves into the calcium shells of dead molluscs, for example, mussels, snails, and oysters (f). Until 1949, it was thought that this stage, which grows right in the shells, was a completely distinct species (*Conchocelis rosea*). Haploid conchospores are formed in this stage, released (g), and deposited on a suitable substrate. They develop into the large-bladed, sexually differentiated generation that is used to make *nori* and the cycle comes full circle. Without detailed knowledge of the *Conchocelis* stage, the efficient commercial production of *Porphyra* would not have been made possible.

Technical and scientific details

Before the British alga researcher Kathleen Mary Drew-Baker identified the *Conchocelis* stage, no one knew where the spores for the fully grown *Porphyra* originated. This was the main reason for the recurring problems experienced by the Japanese seaweed fishers in their attempts to cultivate *Porphyra* in a predictable manner. Drew-Baker was unaware of their difficulties. Instead, she was preoccupied with shedding light on the mystery of why the species of laver (*Porphyra umbilicalis*) that grew around the coast of England seemed to disappear during the summer, reappearing again only toward the end of autumn. She tried without success to germinate spores that she had collected. Finally, after nine years' worth of effort to grow the alga in light- and temperature-controlled tanks, she discovered that the spores would germinate if they were allowed to settle on a sterilized oyster shell. They would even grow on an eggshell. A few months later, these small, roseate sprouts produced their own spores that, in turn, could germinate and develop into the well-known large purple laver.

Drew-Baker published her results in 1949. Shortly thereafter the Japanese phycologist Sokichi Segawa repeated her experiments using local varieties of *Porphyra* and found that they behaved in the same way as the English species. The mystery was solved and the results were quickly put to use in Japan. Already by 1953, the Japanese marine biologist Fusao Ota had thought out a simple way of seeding *Porphyra* spores onto nets that could be suspended in the ocean. The method consisted of allowing the mature *Porphyra* blades to shed their spores in tanks that were filled with discarded oyster shells that, coincidentally, were left over from the harvesting of cultured pearls. When these spores had sprouted and developed into *Conchocelis*, the contents of the tanks were stirred so that the spores from the *Conchocelis* themselves were shaken loose. These spores were able to fasten onto, and sprout on, the strings of nets that were lowered into the tanks. The seeded nets could then be placed in the sea and the cultivation of the much sought-after *nori* could begin.

Kathleen Drew-Baker died at a relatively young age in 1956, apparently unaware that her curiosity and seminal research had laid the foundations for the development of the most valuable aquaculture industry in the world.

▸▸ *Porphyra umbilicalis*, specimen from the private collection of Kathleen Mary Drew-Baker, now housed in the Natural History Museum in London.

Rhosneigr. Anglesey.
11. XII. 1950

FLORA OF BRITISH ISLES

Porphyra umbilicalis

Rhosneigr, Anglesey.

11. 12. 50.

Coll. K. M. Drew   No. 2109b

The nutritional content of seaweeds

*Technical and
scientific details*

The nutritional content of seaweeds cannot be quantified with great precision. This is because their composition in terms of calories, proteins, fats, minerals, vitamins, and dietary fiber depends largely on the individual species, local growing conditions, time of year, and how they have been stored, as well as the means by which they are processed and preserved, if applicable. For the same reasons it is difficult to make recommendations concerning the optimal daily intake.

The tables that follow incorporate data from a number of sources. Some of the information is gleaned from the book *Low Calorie, High Nutrition Vegetables from the Sea* by Seibin and Teruku Arasaki. Even though this volume dates from 1983, it is still considered to be a very authoritative work on seaweeds and their nutritional content. In view of the approximate nature of most of the values reported, the numbers are generally rounded off and refer to the dry weight.

For more information about the chemical composition of species other than those mentioned in the following tables, the reader can consult the scientific literature in the Bibliography below.

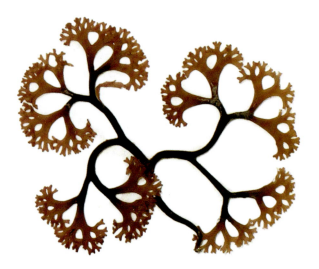

▸ Dried Irish moss
(*Chondrus crispus*).

Iodine and minerals

	I mg/100g	Ca mg/100g	Na mg/100g	K mg/100g	Mg mg/100g	P mg/100g
Winged kelp (*Alaria esculenta*)	17	1,000	4,200	7,500	900	500
Konbu (*Saccharina japonica*)	150–500	800	3,100	6,000	120	150
Hijiki (*Sargassum fusiforme*)	40–60	1,400		15,000		60
Wakame (*Undaria pinnatifida*)	5–40	1,300	1,100	2,000	1,100	260
Bladder wrack (*Fucus vesiculosus*)	30–60	2,200	2,700	2,.00	700	100
Knotted wrack (*Ascophyllum nodosum*)	150	4,200	3,700	2,300	600	200
Arame (*Eisenia bicyclis*)	50–500	1,200		4,000		
Dulse (*Palmaria palmata*)	5	200	1,700	8,000	300	400
Laver, *nori* (*Porphyra* spp.)	0.5–1.5	470	3,600	3,500	600	400
Carrageen (*Chondrus crispus*)	20–30	900	2,200	3,400	700	800
Sea lettuce (*Ulva* spp.)	20	730	1,100	700	2,800	
Green string lettuce (*Ulva* spp.)	7	600	7,500	700	2,700	
Spirulina	0.1	700	900	1,200	180	120
Recommended daily intake	0.15mg	1,000mg	2,400mg	3,500mg	400mg	1,000mg

I = iodine; Ca = calcium; Na = sodium; K = potassium; Mg = magnesium; P = phosphorous
mg = milligram = 0.001 gram

The nutritional content of seaweeds

Sources: Arasaki, S. & T. Arasaki. *Low Calorie, High Nutrition Vegetables From the Sea.* Japan Publications, Inc., Tokyo, 1983, pp. 32–52; Erhart, S. & L. Cerier. *Sea Vegetable Celebration.* Book Publishing Company, Summertown, Tennessee, 2001, pp. 154–155; Teas, J., S. Pino, A. Critchley & L. E. Braverman. Variability of iodine content in common commercially available edible seaweeds. *Thyroid* 14, 836–841, 2004; Rupérez, P. Mineral content of edible marine seaweeds. *Food Chemistry* 79, 23–26, 2002; van Netten, C., S. A. Hoption Cann, D. R. Morley & J. P. van Netten. Elemental and radioactive analysis of commercially available seaweed. *Sci. Total Environ.* 255, 169–175, 2000; www.nal.usda.gov.

TRACE ELEMENTS

	Fe mg/100g	Mn mg/100g	Cu mg/100g	Cr mg/100g	Zn mg/100g	Se mg/100g
Winged kelp (*Alaria esculenta*)	18	1.0	0.17	0.21	3.4	
Konbu (*Saccharina japonica*)	15	1.7	0.5		1	0.0007
Hijiki (*Sargassum fusiforme*)	30					
Wakame (*Undaria pinnatifida*)	13	5	< 0.5		2	0.004
Bladder wrack (*Fucus vesiculosus*)	36	7	0.1	0.2	1.3	
Knotted wrack (*Ascophyllum nodosum*)	50	23	0.2	0.06	1	
Arame (*Eisenia bicyclis*)	12		0.3			
Dulse (*Palmaria palmata*)	33	1.1	0.38	0.15	2.9	
Laver, *nori* (*Porphyra* spp.)	23	2	< 0.5		2	0.001
Carrageen (*Chondrus crispus*)	20	1.3	0.8		8	0.004
Sea lettuce (*Ulva* spp.)	100	1–35				
Green string lettuce (*Ulva* spp.)	15	0.5				
Spirulina	100	2	6	0.2	2	0.006
Recommended daily intake	18mg	2mg	2mg	0.012mg	15mg	0.05mg

Fe = iron; Mn = manganese; Cu = copper; Cr = chromium; Zn = zinc; Se = selenium
mg = milligram = 0.001 gram

Sources: Arasaki, S. & T. Arasaki. *Low Calorie, High Nutrition Vegetables From the Sea.* Japan Publications, Inc., Tokyo, 1983, pp. 32–52; Erhart, S. & L. Cerier. *Sea Vegetable Celebration.* Book Publishing Company, Summertown, Tennessee, 2001, pp. 154–155; Rupérez, P. Mineral content of edible marine seaweeds. *Food Chemistry* 79, 23–26, 2002; van Netten, C., S. A. Hoption Cann, D. R. Morley & J. P. van Netten. Elemental and radioactive analysis of commercially available seaweed. *Sci. Total Environ.* 255, 169–175, 2000; www.nal.usda.gov.

ENERGY CONTENT, FATS, PROTEINS, CARBOHYDRATES, AND FIBER

	Energy content kJ/100g	Fats g/100g	Proteins g/100g	Carbo-hydrates g/100g	Fiber g/100g
Winged kelp (*Alaria esculenta*)	1,100	3.6	18	40	39
Konbu (*Saccharina japonica*)	1,100	1.1	7	55	3
Hijiki (*Sargassum fusiforme*)	1,000	0.8	10	47	17
Wakame (*Undaria pinnatifida*)	1,150	4.5	13	51	4
Bladder wrack (*Fucus vesiculosus*)	1,200	3.6	6	55	4
Knotted wrack (*Ascophyllum nodosum*)	1,000	1	10	52	6
Arame (*Eisenia bicyclis*)	1,100	1.3	12	52	7
Dulse (*Palmaria palmata*)	1,100	1.7	22	45	33
Laver, *nori* (*Porphyra* spp.)	1,350	2	34	45	7
Carrageen (*Chondrus crispus*)	1,050	1–3	15	60	6
Sea lettuce (*Ulva* spp.)	900	0.6	24	40	5
Green string lettuce (*Ulva* spp.)		0.5	21	50	7
Spirulina	1,250	7	65	14	6
Recommended daily intake	8,400+ kJ	65+ g	50+ g	300+ g	25+ g

The nutritional content of seaweeds

mg = milligram = 0.001 gram; kJ = kilojoule = 0.24 kcal

Sources: Arasaki, S. & T. Arasaki. *Low Calorie, High Nutrition Vegetables From the Sea*. Japan Publications, Inc., Tokyo, 1983, pp. 32–52. Erhart, S. & L. Cerier. *Sea Vegetable Celebration*. Book Publishing Company, Summertown, Tennessee, 2001, pp. 154–155; www.nutritiondata.com; www.nal.usda.gov; Dawczynski, C., R. Schubert & G. Jahreis. Amino acids, fatty acids, and dietary fiber in edible seaweed products. *Food Chemistry* 103, 891–899, 2006.

VITAMINS

	A	B_1	B_2	B_3	B_6	B_{12}	C	E
Winged kelp (*Alaria esculenta*)	8,500	0.6	2.7	11	6	5	6	5
Konbu (*Saccharina japonica*)	430	0.08	0.3	2	0.3	0.3	11	
Hijiki (*Sargassum fusiforme*)	150	0.01	0.2	5		0.6	0	4
Wakame (*Undaria pinnatifida*)	140	0.2	0.2	10	0.01	0.6	15	4
Bladder wrack (*Fucus vesiculosus*)								
Knotted wrack (*Ascophyllum nodosum*)			0.5	18			38	
Arame (*Eisenia bicyclis*)	50	0.2	0.2	3			0	
Dulse (*Palmaria palmata*)	660	0.07	2	2	9	7	6	2
Laver, *nori* (*Porphyra* spp.)	30,000	0.2	1	3	1	20	20	
Carrageen (*Chondrus crispus*)	600	0.08	2.5	3	0.4	0	15	4
Sea lettuce (*Ulva* spp.)	1,000	0.06	0.03	8		6	10	
Green string lettuce (*Ulva* spp.)	500	0.04	0.5	1		1.3	10	
Spirulina	550	0.3	3.5	15	0.8	0	9	5
Recommended daily intake	5,000IU	1.5mg	1.7mg	20mg	2mg	6mg	60mg	30IU

Quantities for vitamins A and E are in units of IU/100grams and for vitamins B and C in units of mg/100 grams.

IU = International Units; mg = milligram = 0.001 gram
A: beta-carotene; B_1: thiamine; B_2: riboflavin; B_3: niacin; B_6: pyridoxine; B_{12}: cyanocobalamin

Sources: Arasaki, S. & T. Arasaki. *Low Calorie, High Nutrition Vegetables From the Sea.* Japan Publications, Inc., Tokyo, 1983, pp. 32–52; Erhart, S. & L. Cerier. *Sea Vegetable Celebration.* Book Publishing Company, Summertown, Tennessee, 2001, pp. 154–155; www.nal.usda.gov.

Brown algae
Fatty-acid composition (% of total fat content)

	Alaria	Laminaria	Hijiki	Undaria	Nereocystis	Macrocystis	Fucus
14:0	3.5	2.9	0.3	2.3	9	8	9
16:0	15	36	27	14	15	16	18
18:1	11	13	7.7	6.0	14	12	24
18:2 ω-6	3.7	5.5	3.6	7.4	6	4.3	11
18:3 ω-3	8.7	1.6	0.4	1.7	8	7	5
18:4 ω-3	20	1.2	-	26	12	16	5
20:4 ω-6 (AA)	14	12	5.3	13	17	14	12
20:5 ω-3 (EPA)	16	16	42	13	13	9	6
22:6 ω-3 (DHA)	-	-	-	-	-	-	-
Σsaturated	20	42	28	18	27	26	30
Σmonounsaturated	13	17	13	8	16	15	28
Σpolyunsaturated	65	40	57	74	57	53	42
Σω-3	45	18	43	50	36	31	17
Σω-6	20	22	14	23	24	22	24
Σω-3/ Σω-6	2.3	0.85	3.2	2.2	1.5	1.4	0.7
DHA/AA	-	-	-	-	-	-	-
EPA/AA	1.1	1.3	8	1	0.8	0.6	0.6

The nutritional content of seaweeds

The fatty acids are denoted in the table as n:m, where n is the number of carbon atoms and m is the number of double bonds. ω-6 and ω-3 indicate the type of essential fatty acid. AA = arachidonic acid; EPA = eicosapentaenoic acid; DHA = docosahexaenoic acid. Σ stands for the sum of a given type of fatty acid.

Sources: Dawczynski, C., R. Schubert & G. Jahreis. Amino acids, fatty acids, and dietary fiber in edible seaweed products. *Food Chemistry* 103, 891–899, 2006; Colombo, M. L., P. Risé, F. Giavarini, L. de Angelis, C. Galli & C. L. Bolis. Marine macroalgae as sources of polyunsaturated fatty acids. *Plant Foods Human. Nutr.* 61, 67–72, 2006; Khotimchenko, S. V., V. E. Vaskovsky & T. V. Titlyanova. Fatty acids of marina algae from the Pacific Coast of North California. *Botanica Marina* 45, 17–22, 2002.

RED ALGAE, GREEN ALGAE, AND SPIRULINA
FATTY-ACID COMPOSITION (% OF TOTAL FAT CONTENT)

	Porphyra	Palmaria	Ulva	Spirulina
14:0	3	8	0.4	0.4
16:0	33	27	24	27
16:4 ω-3	-		14	-
18:1	11	9	1.4	34
18:2ω-6	5		4	12
18:3 ω-3	3		22	0.7
18:4 ω-3	3		12	0.7
20:4 ω-6 (AA)	9	5	0.5	0.4
20:5 ω-3 (EPA)	15	34	1.4	2.5
22:6 ω-3 (DHA)	-	-	-	3
Σsaturated	40	43	26	35
Σmonounsaturated	20	14	3	38
Σpolysaturated	38	43	57	24
Σω-3	20	35	54	7
Σω-6	17	8	3.3	17
Σω-3/ Σω-6	2	4	16	0.3
DHA/AA	-	-	-	2.3
EPA/AA	2	7	3	6

The fatty acids are denoted in the table as n:m, where n is the number of carbon atoms and m is the number of double bonds. ω-6 and ω-3 indicate the type of essential fatty acid. AA = arachidonic acid; EPA = eicosapentaenoic acid; DHA = docosahexaenoic acid. Σ stands for the sum of a given type of fatty acid.

Sources: Dawczynski, C., R. Schubert & G. Jahreis. Amino acids, fatty acids, and dietary fiber in edible seaweed products. *Food Chemistry* 103, 891–899, 2006; Colombo, M. L., P. Risé, F. Giavarini, L. de Angelis, C. Galli & C. L. Bolis. Marine macroalgae as sources of polyunsaturated fatty acids. *Plant Foods Human. Nutr.* 61, 67–72, 2006; Khotimchenko, S. V., V. E. Vaskovsky & T. V. Titlyanova. Fatty acids of marine algae from the Pacific Coast of North California. *Botanica Marina* 45, 17–22, 2002; Tokuşoglu, Ö. & M. K. Ünal. Biomass nutrient profiles of three microalgae: *Spirulina platensis, Chlorella vulgaris,* and *Isochrisis galbana. J. Food Science* 68, 1144–1148, 2003; Professor Gerhard Jahreis, Friedrich-Schiller University, Jena, has kindly analyzed the fat content in *Palmaria* in a sample from Mendocino, California.

POLYSACCHARIDES IN SEAWEEDS—ALGINATES, CARRAGEENANS, AND AGARS

Both plants and macroalgae make use of polysaccharides to act as energy depots and as structural elements in cell walls and to build larger structures, such as stems and leaves and stipes and blades, respectively. In contrast to the simpler and well-defined polysaccharides used by plants to store energy, for example, glycogen and starch, those utilized by algae for structural purposes are more complex and heterogeneous. Plants avail themselves of a class of water soluble polysaccharides called pectins, which are often used to help set jams and jellies. Seaweeds, on the other hand, contain polysaccharides that are unique to them, namely, alginates, carrageenans, and agars, which are all soluble dietary fiber.

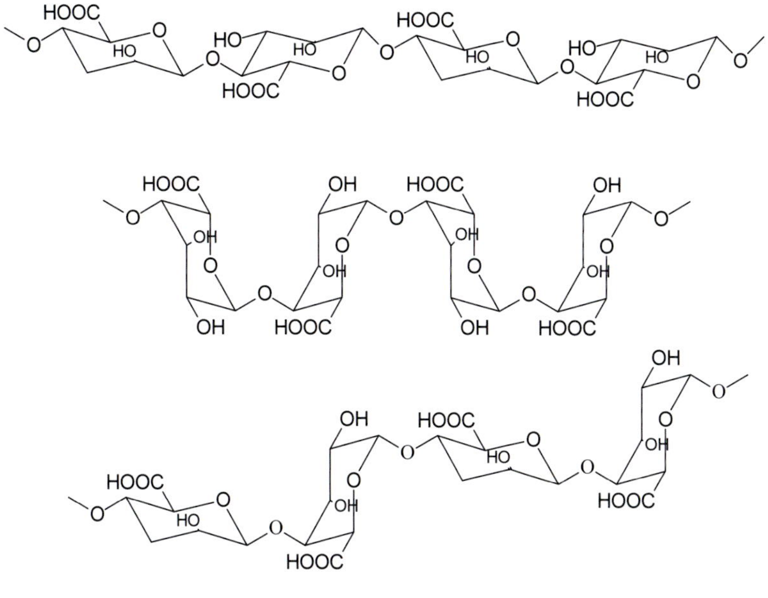

The nutritional content of seaweeds

◄ Alginates are polysaccharides formed from long chains of two different types of monosaccharide groups, β-D-mannuronic acid (M) and α-L-guluronic acid (G).

Alginates are found in the brown macroalgae. They are made up of long, linear molecules, which are composed of two different monosaccharide groups, β-D-mannuronic acid (M) and α-L-guluronic acid (G). As shown in the illustration above, these acid groups can be bound linearly, for example, as -M-M-M-M-, -G-G-G-G-, or -M-G-M-G-.

Alginate is the basic form of these substances and their complementary ammonium and alkali salts, for example, sodium alginate, are water soluble. The content of the acids M and G in an alginate is dependent on the seaweed species and the part of the organism from which it is extracted. Chain length also varies, with the shortest typically composed of 500 monosaccharide groups.

Technical and scientific details

The usefulness of alginates, particularly in the form of sodium alginate, rests on their water solubility. In an aqueous solution, sodium alginate is found in the ionic state as a so-called electrolyte.

Alginates form gels in the presence of Ca^{++} (or other divalent ions such as Mg^{++} and Ba^{++}), a process that occurs at much lower temperatures than is possible with pectins. Gelation with Ca^{++} provides for a stronger gel, whereas Mg^{++} leads to a weak gel, both with different rheological properties and hence different applications. The affinity for Ca^{++}, which is abundant in milk, also means that combinations of alginate and milk are particularly useful.

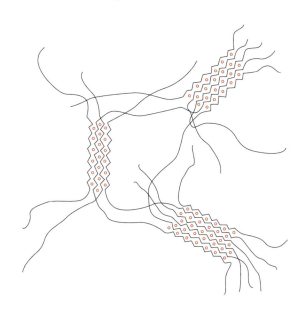

▶ Alginates can form gels in water when the long polysaccharide chains are bound together by Ca^{++} ions (shown in red).

▶ Some agars and carrageenans in an aqueous solution form double helices when it is cooled. These can bind to each other in a variety of ways to form networks (gels). Carrageenans are differentiated from each other by their ability and tendency to form different types of gels and molecular aggregates.

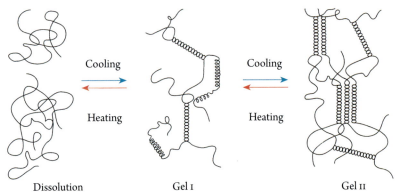

Dissolution   Cooling → / ← Heating   Gel I   Cooling → / ← Heating   Gel II

248

The melting point of alginate gels lies just above the boiling point of water. In the course of gelation, large quantities of water are bound, which is the basis for the application of alginates as thickeners and stabilizers. In addition, these gels are acid resistant, which gives them an advantage over other types of stabilizers.

The polysaccharides carrageenan and agar, which are found in red algae, have a somewhat more complex structure than alginate. They are composed of the monosaccharide galactose in chemical compounds with differing degrees of sulphation.

Carrageenans are long, flexible chains made up of about 25,000 galactose sub-units. Individual carrageenans have different properties with respect to gelation and their behavior is dependent on factors such as pH, ion content, and temperature. Carrageenans are polyelectrolytes, with the ability to form gels in the presence of K^+ and Ca^{++} ions. Some can form double helices that, in turn, are able to link loosely together to make a network. There are three important carrageenans that have technical end uses: κ-carrageenan, which forms strong, stiff gels; ι-carrageenan, which forms soft gels; λ-carrageenan, which is well suited for emulsifying proteins. Only the last one is soluble in cold water.

Agar, known also as agar-agar and *kanten*, consists of large molecules that are composed of two types of galactose sub-units, agarose and agaropectin, which differ with respect to their sulphate content. Like carrageenan, agar can form double helices and networks. In this way, fibers can be created that are made up of thousands of chains. Agar is insoluble in cold water, but dissolves readily in boiling water. It has a formidable capacity for forming gels that contain up to 99.5% water and are stable up to 85°C, their melting point.

Because agar, carrageenan, and alginate are long-chained molecules they impart interesting fluid properties to aqueous solutions. On the one hand, they make the solutions more viscous and stiff. On the other, they help to make the solutions flow more easily when subjected to streaming because the long molecules align themselves so that they can more readily slip past each other. Fluids of this type, which are familiar to us from the example of ketchup, are called complex fluids or non-Newtonian fluids. When the ketchup bottle is stood on its head, the ketchup, which is very viscous, flows out really slowly. But if one shakes the bottle, the ketchup gains speed and starts to flow more readily, because the viscosity is reduced as a result of streaming. This effect is known as 'shear-thinning'.

The nutritional content of seaweeds

ι-carrageenan (iota)

κ-carrageenan (kappa)

λ-carrageenan (lamda)

▲ Three different types of carrageenan.

Names of the algae, seaweeds, and marine plants in the book

Technical and scientific details

Some of the seaweed species discussed in this book have no vernacular name in English. In other cases, they have more than one, or the same name is used interchangeably for several species. A classical illustration of the latter is the word 'kelp'.

Biologists use unique Latin designations, consisting of the name of the *genus*, always given first and capitalized, followed by that of the *species*. Sometimes, the name of the individual who first described and named the organism and the year he or she did so is tagged on. (An example is *Macrocystis integrifolia* Bory de Saint-Vincent 1826.) The designation of a species can change when further research—for instance, re-interpretation of its genetic information or discoveries about complicated life cycles—reveals new aspects of an organism that differentiate it from the others. For example, it has recently been suggested that some members of the genus *Porphyra* should be reclassified in a whole new genus called *Pyropia*.

When referring either to one or to several unspecified species from the same *genus*, one writes *Genus* sp. or *Genus* spp., respectively.

The lists below include many of the seaweed and marine plant species mentioned in this book. Note that sometimes there is no English equivalent for the common Japanese name. In the case of others that have been reclassified, the older designation, still in use, is given in brackets.

COMMON ENGLISH NAMES

arame—*Eisenia bicyclis* or *Eisenia arborea*
badderlocks—*Alaria esculenta*
black carrageen—*Furcellaria lumbricalis*
bladder wrack—*Fucus vesiculosus*
branched string lettuce—*Ulva prolifera*
brown algae—Phaeophyceae
bullwhip kelp—*Nereocystis luetkeana*—also
  bull kelp, ribbon kelp
carrageen—*Chondrus crispus*
cuvie—*Laminaria hyperborea*
dabberlocks—*Alaria esculenta*
dulse—*Palmaria palmata*
eelgrass—*Zostera marina*
giant kelp—*Macrocystis pyrifera*
glasswort—*Salicornia europaea*
green algae—Chlorophyta
green string lettuce—*Ulva linza*
gulfweed—*Sargassum*

hijiki—*Sargassum fusiforme*
Irish moss—*Chondrus crispus*
konbu—(*kombu*) primarily *Saccharina japonica*
knotted wrack—*Ascophyllum nodosum*
laver—*Porphyra* spp.
oarweed—*Laminaria digitata*
purple laver—*Porphyra umbilicalis*
red algae—Rhodophyta
ribbon kelp—*Alaria marginata*
rock weed—*Ascophyllum nodosum*
samphire—*Salicornia europaea*
sea asparagus—*Salicornia europaea*
sea beech—*Delesseria sanguinea*
sea lettuce—*Ulva lactuca*
sea-girdles—*Laminaria digitata*
sea grapes—*Caulerpa lentillifera*
sea moss—*Gracilaria* spp.
sea oak—*Halidrys siliquosa*

sea palm—*Postelsia palmaeformis*
sea pickle—*Salicornia europaea*
sea tangle—*Laminaria* spp.
sea thong—*Himanthalia lorea*
serrated wrack—*Fucus serratus*
stringy acid kelp—*Desmarestia viridis*
sugar kelp—*Laminaria saccharina, Saccharina latissima*

sugar wrack—*Laminaria saccharina, Saccharina latissima*
tangle—*Laminaria digitata*
tangleweed—*Laminaria hyperborea*
thongweed—*Himanthalia elongata*
wakame—*Undaria pinnatifida*
wild Atlantic *konbu*—*Laminaria longicruris*
winged kelp—*Alaria esculenta*

BROWN MACROALGAE (PHAEOPHYCEAE)

Alaria esculenta winged kelp, badderlocks, dabberlocks (far North Atlantic area)
Alaria marginata ribbon kelp (North Pacific)
Ascophyllum nodosum knotted (egg) wrack
Cladosiphon okamuranus *mozuku* (Japan)
Desmarestia viridis stringy acid kelp
Eisenia arborea *arame*
Eisenia bicyclis *arame*
Fucus evanescens no common name in English
Fucus vesiculosus bladder wrack
Fucus serratus serrated wrack
Halidrys siliquosa sea oak
Himanthalia elongata thongweed
Himanthalia lorea sea thong
Laminaria digitata oarweed, sea-girdles, tangle

Laminaria hyperborea tangleweed, cuvie
Laminaria setchellii wild North Pacific *konbu*
Macrocystis pyrifera (*Macrocystis integrifolia*) giant kelp, giant perennial kelp, iodine kelp, long bladder kelp
Nereocystis luetkeana bullwhip kelp, bull kelp
Postelsia palmaeformis sea palm
Saccharina japonica (*Laminaria japonica*) *konbu*
Saccharina latissima (*Laminaria saccharina*) sugar kelp, sugar wrack
Saccharina longicruris (*Laminaria longicruris*) wild Atlantic *konbu*
Sargassum fusiforme (*Hizikia fusiforme*) *hijiki*
Sargassum spp. gulfweed
Undaria pinnatifida *wakame*

RED MACROALGAE (RHODOPHYTA)

Chondrus crispus carrageen, Irish moss
Conchocelis rosea stage of the *Porphyra* life cycle
Delesseria sanguinea sea beech
Furcellaria lumbricalis (*Furcellaria fastigiata*) black carrageen
Gelidium amansii *tengusa*
Gracilaria verrucosa sea moss, *ogonori*
Meristotheca papulosa *tosaka-nori* (Japan)

Palmaria palmata (*Rhodymenia palmata*) dulse
Porphyra abbottiae laver
Porphyra nereocystis laver (North Pacific)
Porphyra perforata laver (North Pacific)
Porphyra tenera *nori* (Japan)
Porphyra umbilicalis purple laver (North Atlantic)
Porphyra yezoensis *nori* (Japan)

GREEN MACROALGAE (CHLOROPHYTA)

Caulerpa lentillifera sea grapes
Monostroma latissimum green alga, *hiteogusa* (Japan)
Monostroma nitidum green alga, *hiteogusa* (Japan)

Ulva lactuca sea lettuce
Ulva linza green string lettuce
Ulva prolifera branched string lettuce
Ulva spp. (*Enteromorpha* spp.) sea lettuces

BLUE-GREEN MICROALGAE (CYANOBACTERIA)

Arthrospira maxima Spirulina

Arthrospira platensis Spirulina

EUKARYOTIC MICROALGAE

Chlorella spp. chlorella, green microalga

MARINE PLANTS

Salicornia europaea sea asparagus, samphire, glasswort, sea pickle

Zostera marina eelgrass

Bibliography

SEAWEEDS IN COOKING

Andoh, E. *Washoku. Recipes From the Japanese Home Kitchen.* Ten Speed Press, Berkeley, 2005.

Arasaki, S. & T. Arasaki. *Low Calorie, High Nutrition Vegetables From the Sea.* Japan Publications, Inc., Tokyo, 1983.

Babel, K. *Seafood Sense.* Basic Health Publ., Laguna Beach, California, 2005.

Blumenthal, H., P. Barbot, N. Matsushisa & K. Mikuni. *Dashi and Umami. The Heart of Japanese Cuisine.* Eat-Japan, Cross Media Ltd., London, 2009.

Bradford, P. & M. Bradford. *Cooking with Sea Vegetables.* Healing Arts Press, Rochester, Vermont, 1985.

Chavannes, C. D. *Algues. Légumes de la mer.* La Plage Ed., Sète, 2002.

Cooksley, V. G. *Seaweed. Nature's Secret to Balancing Your Metabolism, Fighting Disease, and Revitalizing Body & Soul.* Stewart, Tabori & Chang, New York, 2007.

Dubin, M. D. & S. Ross. *Seaweed, Salmon, and Manzanita Cider: A California Indian Feast.* Heyday, Berkeley, 2008.

Ellis, L. *Seaweed. A Cook's Guide.* Fisher Books, LCC, Tuscon, Arizona, 1998.

Erhart, S. & L. Cerier. *Sea Vegetable Celebration.* Book Publishing Company, Summertown, Tennessee, 2001.

Fryer, L. & D. Simmons. *Food Power From the Sea. The Seaweed Story.* Acres USA, Austin, Texas, 1977.

Fujii, M. *The Enlightened Kitchen.* Kodansha International, New York, 2005.

Gusman, J. *Vegetables From the Sea. Everyday Cooking With Sea Greens.* William Morrow, HarperCollinsPublishers, New York, 2003.

Harbo, R. M. *The Edible Seashore. Pacific Shores Cookbook & Guide.* Hancock House Publ. Ltd., Surrey, British Columbia, Canada, 1988.

Huston, F. & X. Milne. *Seaweed and Eat It. A Family Foraging and Cooking Adventure.* Virgin Books Ltd., London, 2008.

Lewallen, E. & J. Lewallen. *Sea Vegetable Gourmet Cookbook and Wildcrafter's Guide.* Mendocino Sea Vegetable Co., Mendocino, California, 1996.

Maderia, C. J. *The New Seaweed Cookbook.* North Atlantic Books, Berkeley, CA, 2007.

Madlener, J. C. *The Sea Vegetable Book.* Clark-son N. Potter, Inc., Publ., New York, 1977.

Madlener, J. C. *The Sea Vegetable Gelatin Cookbook and Field Guide.* Woodbridge Press Publ. Co., Santa Barbara, CA, 1981.

McConnaughey, E. *Sea Vegetables. Harvesting Guide & Cookbook.* Naturegraph Publishers, Inc., Happy Camp, California, 2002.

Mouritsen, O. G. *Sushi. Food for the Eye, the Body & the Soul*, Springer, New York, 2009.

Mouritsen, O. G., L. Williams, R. Bjerregaard & L. Duelund. Seaweeds for umami flavor in the New Nordic Cuisine. *Flavour* 1:4, 2012.

Mouritsen, O. G. The emerging science of gastrophysics and its application to the algal cuisine. *Flavour* 1:6, 2012.

Rhatigan, P. *Irish Seaweed Kitchen.* Booklink, Co Down, Ireland, 2009.

Shimbo, H. *The Sushi Experience.* Alfred A. Knopf, New York, 2006.

Turner, N. J. *Plants of Haida Gwaii.* Sonosis Press, Winlaw, British Columbia, 2004.

Turner, N. J. *Food Plants of Coastal First Peoples.* UBC Press, Vancouver, 2006.

SEAWEEDS AND ALGAE

Barsanti, L. & P. Gualtieri. *Algae. Anatomy, Biochemistry, and Biotechnology.* CRC Press, Taylor & Francis, Boca Raton, 2006.

Becker, H. *Seaweed Memories. In the Jaws of the Sea.* Wolfhound Press, Dublin, 2002.

Braune, W. & M. D. Guiry. *Seaweeds.* A. R. G. Gartner Verlag KG, Ruggell, Germany, 2011.

Chapman, V. J. *Seaweeds and Their Uses.* Methuen & Co. Ltd., London, 1950.

Critchley, A. T., M. Ohno & D. B. Largo. *World Seaweed Ressources. An Authoritative Reference System.* Ver. 1.0, DVD ROM, ETI Inf. Services Ltd, Workingham, UK, 2006.

Crowe, A. *A Field Guide to the Native Edible Plants of New Zealand.* Penguin Books, Auckland, 2006.

Daegling, M. *Monster Seaweeds. The Story of the Giant Kelps.* Dillon Press Inc., Minneapolis, 1986.

Davenport, J., K. Black, G. Burnell, T. Cross, S. Culloty, S. Ekaratne, B. Furness, M. Mulcahy & H. Thetmeyer. *Aquaculture.* Blackwell Publ. Co., Oxford, UK, 2003.

Druehl, L. *Pacific Seaweeds. A Guide to Common Seaweeds of the West Coast.* Harbour Publishing, British Columbia, Canada, 2000.

O'Clair, R. M. & S. C. Lindstrom. *North Pacific Seaweeds*. Plant Press, Auke Bay, AK, 2000.

Pedersen, P. M. *Grønlands Havalger*. Epsilon, Copenhagen, 2011.

Tilden, J. E. T*he Algae and Their Life Relations*. Univ. Minnesota Press., Minneapolis, 1935.

The State of World Fisheries and Aquaculture, FAO; 2010 [http://www.fao.org/docrep/013/i1820e/-i1820e00.htm]

Thomas, D. *Seaweeds*. The Natural History Museum, Life Series, London, 2002.

Wynne, M. J. *Portraits of Marine Algae: an Historical Perspective*. University of Michigan Herbarium, Ann Arbor, Michigan, 2006.

SCIENTIFIC LITERATURE

Abdussalam, S. Drugs from seaweeds. *Medical Hypotheses* 32, 33–35, 1990.

Almela, C., S. Algora, V. Benito, M. J. Clemente, V. Devesa, M. A. Suner, D. Velez, & R. Montoro. Heavy metal, total arsenic, and inorganic arsenic contents of algae food products. *J. Agric. Food Chem.* 50, 918–923, 2002

Andersen, R. A. Diversity of eukaryotic algae. *Biodiversity and Conservation* 1, 267–292, 1992.

Bartie, W., P. Madorin & G. Ferlan. Seaweed, vitamin K, and warfarin. *Amer. J. Health Syst. Pharm.* 58, 2300, 2001.

Burtin, P. Nutritional value of seaweeds. *Electron. J. Environ. Agric. Food Chem.* 2, 498–503, 2003.

Bixler, H. & H. Porse. A decade of change in the seaweed hydrocolloid industry. *J. Appl. Phycol.* 23, 321–335, 2011.

Colombo, M. L., P. Risé, F. Giavarini, L. de Angelis, C. Galli & C. L. Bolis. Marine macroalgae as sources of polyunsaturated fatty acids. *Plant Foods Human. Nutr.* 61, 67–72, 2006.

Cornish, M. L. & D. J. Garbary. Antoioxidants from macroalgae: potential applications in human health and nutrition. *Algae* 25, 155–171, 2010.

Crawford, C. & D. Marsh. *The Driving Force. Food, Evolution, and the Future*. Harper & Row, Publishers, London, 1989.

Cunnane, S. C. *Survival of the Fattest*. World Scientific, London, 2005.

Cunnane, S. C. & K. M. Stewart. *Human Brain Evolution. The Influence of Freshwater and Marine Food Resources*. Wiley-Blackwell, New Jersey, 2010.

Dam, H. & J. Glavind. Vitamin K in the plant. *Biochem. J.* 32, 485–487, 1938.

Dawczynski, C., R. Schubert & G. Jahreis. Amino acids, fatty acids, and dietary fiber in edible seaweed products. *Food Chem.* 103, 891–899, 2006.

Dillehay, T. D., C. Ramírez, M. Pino, M. B. Collins, J. Rossen & J. D. Pino-Navarro. Monte Verde: Seaweed, food, medicine, and the peopling of South America. *Science* 320, 784–786, 2008.

Duarte, C. M., N. Marbá & M. Holmer. Rapid domestication of marine species. *Science* 316, 382–383, 2007.

Gerschwin, M. E. & A. Belay (eds.). *Spirulina in Human Nutrition and Health*. CRC Press, Boca Raton, 2008.

Hafting, J. T., A. T. Critchley, M. L. C. Scott, A. Hubley & A. F. Archibald. On-land cultivation of functional seaweed products for human usage. *J. Appl. Phycol.* 24:385–392, 2011.

Harding, F. *Breast Cancer: Cause, Prevention, Cure*. Tekline Publ., London, 2007.

Hehemann, J.-H., G. Correc, T. Barbeyron, W. Helbert, M. Czjzek & G. Michel. Transfer of carbohydrate-active enzymes from marine bacteria to Japanese gut microbiota. *Nature* 464, 908–912, 2010.

Holdt, S. & S. Kraan. Bioactive compounds in seaweed; functional food applications and legislation. *J. Appl. Phycol.* 23, 543–597, 2011.

Kovalenko, I., B. Zdyrko, A. Magasinski, B. Hertsberg, Z. Milicev, R. Burtovyy, I. Luzinov & G. Yushin. A major constituent of brown algae for use in high-capacity Li-ion batteries. *Science* 334, 75–79, 2011.

Jensen, G. M., M. Kristensen & A. Astrup. Effect of alginate supplementation on weight loss in obese subjects completing a 12-wk energy-restricted diet: a randomized controlled trial. *Am. J. Clin. Nutr.* 96, 5–13, 2012.

Khotimchenko, S. V., V. E. Vaskovsky & T. V. Titlyanova. Fatty acids of marina algae from the Pacific Coast of North California. *Botanica Marina* 45, 17–22, 2002.

Kupper, F. C., L. J. Carpenter, G. B. McFiggans, C. J. Palmer, T. J. Waite, E. Boneberg, S. Woitsch, M. Weiller, R. Abela, D. Grolimund, P. Potin, A. Butler, G. W. Luther III, P. Kroneck, W. Meyer-Klaucke & M. C. Feiters. Iodide accumulation provides kelp with an inorganic antioxidant impacting atmospheric chemistry. *Proc. Natl. Acad. Sci. USA* 105, 6954–6958, 2008.

Kristjánsson, L. *Íslenzkir Sjávarhættir I*. Bókaútgáfa Menningarsjóðs, Reykjavík, 1980.

Kushi, L. H., J. E. Cunningham, J. R. Hebert, R. H. Lerman, E. V. Bandera & J. Teas. The macrobiotic diet in cancer. *J. Nutr.* 131, 3056S-3064S, 2001.

Lane, C. E., C. Mayes, L. D. Druehl & G. W. Saunders. A multi-gene molecular investigation of the kelp (Laminariales, Phaeophyceae) supports substantial taxonomic re-organization. *J. Phycol.* 42, 493–512, 2006.

Ley, B. M. *Chlorella. The Ultimate Green Food: Nature's Richest Source of Chlorophyll,* DNA *and* RNA. BL Publications, USA, 2003.

Lüning, K. & S. Pang. Mass cultivation of seaweeds: current aspects and approaches. *J. Appl. Phycol.* 15, 115–119, 2003.

MacArtain, P., C. I. Gill, M. Brooks, R. Campbell & I. R. Rowland. Nutritional value of edible seaweeds. *Nutr. Rev.* 65, 535–543, 2007.

Malin, G., S. M. Turner & P. S. Liss. Sulfur: the plankton/climate connection. *J. Phycol.* 28, 590–597, 1992.

Masato, N., S. Hideyuki, T. Itsuko, A. Ritsuo & H. Shysuke. Effect of fucoidan from *Cladisiphon okamuranus* on the eradication of *Helicobacter pylori. Cell* (Tokyo) 37, 452–455, 2005.

McHugh, D. J. A guide to the seaweed industry. *FAO Fisheries Technical Paper.* No. 441, Rome, FAO, 2003.

Miura, A. *Porphyra* cultivation in Japan. In *Advances in Phycology in Japan* (J. Tokida & H. Hirose, eds.). pp. 273–304, 1975.

Mouritsen, O. G. & M. A. Crawford (eds.). Poly-unsaturated fatty acids, neural function and mental health. *Biol. Skr. Dan. Vid. Selsk.* 56, 1–87, 2007.

Mouritsen, O. G., C. Dawczynski, L. Duelund, G. Jahreis, W. Vetter & M. Schröder. On the human consumption of the red seaweed dulse (*Palmaria palmata* (L.) Kuntze). Preprint, 2012.

Ninomiya, K. Umami: a universal taste. *Food Rev. Int.* 18, 23–38, 2002.

Nisizawa, K. *Seaweeds Kaiso. Bountiful Harvest from the Seas.* Japan Seaweed Association, 2002.

Nwosu, F., J. Morris, V. A. Lund, D. Stewart, H. A. Ross & G. J. McDougall. Anti-proliferative and potential anti-diabetic effects of phenolic-rich extracts from edible marine algae. *Food Chem.* 126, 1006–1012, 2011.

Oohusa, T. The cultivation of *Porphyra* "Nori". In *Seaweed Cultivation and Marine Ranching.*

1st Ed. (M. Ohno & A. T. Critchley, eds.). Kanagawa International Fisheries Training Center, Japan International Cooperation Agency, pp. 57–73, 1993.

Pomin, V. M. (ed.). *Seaweed: Ecology, Nutrient Composition, and Medicinal Uses.* Nova Science Publishers, Inc., New York, 2011.

Prasher, S. O., M. Beaugeard, J. Hawari, P. Bera, R. M. Patel & S. M. Kim. Biosorption of heavy metals by red algae (*Palmaria palmata*). *Environ. Technol.* 25, 1097–1106, 2004.

Puch, P.-F. Hominoid lifestyle and diet reconsidered: paleo-environmental and comparative data. *Human Evol.* 15, 151–162, 2000.

Rupérez, P. Mineral content of edible marine seaweeds. *Food Chem.* 79, 23–26, 2002.

Ramsey, U. P., C. J. Bird, P. F. Shacklock, M. V. Laycock & J. L. C. Wright. Kainic acid and 1'-hydroxykainic acid from Palmariales. *Nat. Toxins* 2, 286–292, 1994.

Rosenfeld, L. Discovery and early uses of iodine. *J. Chem. Edu.* 77, 984–987, 2000.

Sánchez-Machado, D. I., J. Lopez-Cervantes, J. López-Hernández & P. Paseiro-Losada. Fatty acids, total lipid, protein, and ash contents of processed edible seaweeds. *Food. Chem.* 85, 439–444, 2004.

Sillen, A. Strontium-calcium ratios (Sr/Ca) of *Australopithecus robustus* and associated fauna from Swartkrans. *J. Human. Evol.* 23, 495–516, 1992.

Simopoulos, A. P. The importance of the ratio of omega-6/omega-3 essential fatty acids. *Biomed. Pharmacotherapy* 56, 365–379, 2002.

Skribola, C. F. The effect of *Fucus vesiculosus*, an edible brown seaweed, upon menstrual cycle length and hormonal status in three pre-menopausal women: a case report. *BMC Complement Altern. Med.* 4, 4–10, 2004.

Smit, A. J. Medicinal and pharmaceutical uses of seaweeds: a review. *J. Appl. Phycol.* 16, 245–262, 2005.

Smyth, P. P. A. The thyroid, iodine, and breast cancer. *Breast Cancer Res.* 5, 235–238, 2003.

Southerland, J. E. *et al.* A new look at an ancient order: genetic revision of Bangiales (Rhodophyta). *J. Phycol.* 47, 1131-1151, 2011.

Tanaka, T., J. Kakino & M. Miyata. Existing conditions and problems on nori (*Porphyra*) cultivation at the coast of Chiba Prefecture in Tokyo Bay. *Nat. Hist. Res.,* Special Issue 3, 97–109, 1997.

Tanaka, Y., S. C. Skoryna & D. Waldron-Edward. Studies on the inhibition of intestinal absorption of radioactive strontium. VI. Alginate degradation products as potent *in vivo* sequestering agents of radioactive strontium. *Can. Med. Assoc. J.* 98, 1179–1182, 1968.

Teas, J. & M. R. Irhimeh. Dietary algae and HIV/AIDS: proof of concept clinical data. *J. Appl. Phycol.* 24, 575–582, 2012.

Teas, J., S. Pino, A. Critchley & L. E. Braverman. Variability of iodine content in common commercially available edible seaweeds. *Thyroid* 14, 836–841, 2004.

Teas, J., J. R. Hebert, J. H. Fitton & P. V. Zimba. Algae—a poor man's HAART? *Medical Hypotheses* 62, 507–510, 2004.

Tobacman, J. K., S. Bhattacharyya, A. Borthakur & P. K. Dudeja. The carrageenan diet: not recommended. *Science* 321, 1040–1041, 2008.

Teas, J., T. G. Hurley, J. R. Hebert, A. A. Franke, D. W. Sepkovic & M. S. Kurzer. Dietary seaweed modifies estrogen and phytoestrogen metabolism in healthy postmenopausal women. *J. Nutr.* 139, 939–944, 2010.

Tokuşoglu, Ö. & M. K. Ünal. Biomass nutrient profiles of three microalgae: *Spirulina platensis, Chlorella vulgaris,* and *Isochrisis galbana. J. Food Science* 68, 1144–1148, 2003.

Valentine, R. C. & D. L. Valentine. *Omega-3 Fatty Acids and the DHA Principle.* CRC Press, Boca Raton, 2010.

van Netten, C., S. A. Hoption Cann, D. R. Morley & J. P. van Netten. Elemental and radioactive analysis of commercially available seaweed. *Sci. Total Environ.* 255, 169–175, 2000.

Zava, T. T., & D. T. Zava. Assessment of Japanese iodine intake based on seaweed consumption in Japan: A literature-based analysis. *Thyroid Res.* 4.14, 2011.

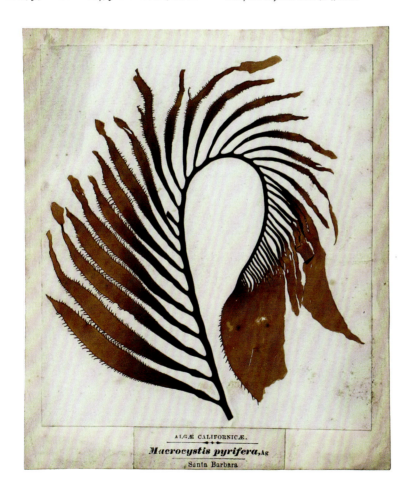

◄ Dried specimen of giant kelp (*Macrocystis pyrifera*). From the collections of the Natural History Museum in London.

Seaweeds on the web

www.algaebase.org the most authoritative, comprehensive, and up-to-date online database on terrestrial and marine algae as well as sea grasses from more than 200 countries. The data base contains information on more than 130,000 species, presented along with illustrations and photos as well as geographical distribution and references to the international literature.

www.isaseaweed.org homepage for *The International Seaweed Organization* that represents interests in research and industrial applications of algae.

www.seaweed.ie a very extensive homepage with general information on algae, aquaculture, and uses of algae as food and medicine. The page has an impressive collection of images of seaweeds.

www.psaalgae.org homepage for *The Phycological Society of America* with comprehensive information and links to museums, collections, and algae by taxa and location around the world.

Technical and scientific details

▶ Production of *nori* from *Porphyra* in Japan. Woodblock print by Katsukawa Shunsen (1762-ca.1830).

Figure credits

Acadian Seaplants: p. 40. bar'sushi: pp. 159, 160. Bech-Poulsen, C.: pp. 71, 91, 97, 125, 155, 181. Claubuesch, S. R.: p. 73. de Zubiria, J. L. L.: p. 195, 197. Fontana, Y.: p. 22. Dillehay, T.: p. 9. Food Snob: 191. Frederiksen, E. and Magasinet Smag & Behag: pp. 116–117, 145. Friðgeirsson, E.: p. 62. Frost, M.: pp. 66–67, 209. G̲iitsx̲aa: p. 133. Hertz, O.: p. 221. Jens Møller Products: p. 179. Kawashima, S.: p. 50. Kikuchi, N.: p. 28. Kinman, N.: pp. 59, 77mr. Kraan, S.: p. 223. Lewis, R.: pp. 202–203. Locher, C., Statens Museum for Kunst KMS1210, SMK Foto: p. 15. Morse, R., Golden State Images, KLG-046, *Macrocystis pyrifera*: pp. x-1; Golden State Images KLG-052: *Macrocystis pyrifera*: p. 3. Mouritsen, J. D.: pp. iv, 19, 23, 41, 47, 48, 49, 65, 68, 69b, 70, 72, 74, 75, 84, 85, 88, 89, 90, 92, 93, 121, 122, 123, 126, 127, 128, 130, 131, 134, 135, 138, 140, 141, 142, 143, 147, 148, 149, 152, 157b, 161, 162, 164, 166, 167, 174, 175, 176, 177, 178, 182, 183, 184, 187b, 189, 232–233, 234, 235, 237, 248, 249. Mouritsen, O. G.: pp. 4, 30, 31, 34, 35, 47, 53, 77, 79, 115, 118, 120, 153, 199, 201, 211, 224, 229, 287. Museum of Science and Industry, Manchester: p. 24. Nielsen, N. E., Læsø Museum: p. 20. Niu, G., Getty Images 81807374: p. 27. Nordic Food Lab (Claes Bech-Poulsen): p. 187t. Notvik, M.: 95. Pedersen, C. Th.: p. 247. Rasmussen, M. B.: p. 214. Ryuusei, M.: p. 25. Saga Prefectural Ariake Fisheries Experimental Station: pp. 43, 57, 58. Selwyn's Penclawdd Seafoods: p. 151. Svensen, E.: 219. Thue, J. and Natural History Museum, London: pp. 5, 13, 15, 19, 39, 83, 86–87, 103, 107, 114, 139, 239, 240, 255, 257, 271. Tsukii, Y.: p. 94. Vea, J.: FMC Biopolymer A/S: p. 207. Williams, L.: 193. World Seaweed Resources: pp. 12, 100. Yamagataya Noriten Limited Company: pp. 157t, 256. Yotsukura, N.: p. 69t.

▶▶ Young fronds of sugar kelp (*Saccharina latissima*). From the collections of the Natural History Museum in London.

Filey. M. E. George.

Glossary

AA see arachidonic acid.

acrylic acid the simplest unsaturated carboxylic acid, found, for example, in brown algae; can have an antibacterial effect.

adenosine triphosphate (ATP) chemical compound (nucleotide) that is a source of energy; together with ADP (adenosine diphosphate) it is involved in virtually all biochemical processes that require energy.

agar polysaccharide, also known as agar-agar, *tengusa*, and *kanten*. Agar is made up of long molecules that are composed of two types of galactose groups, agarose and agaropectin, which contain different sulphate groups. Agar forms double helices and networks in the same way as carrageenan, allowing it to aggregate into fibers with thousands of chains. Agar is insoluble in cold water, but dissolves readily in boiling water. It has the formidable property of being able to gel in a mixture of up to 99.5% water. The gels are stable up to 85°C, its melting point. The strength of the gel is proportional to the sulphate content.

agaropectin polysaccharide; together with agarose makes up agar. Like agarose, it is made up of galactose groups, but has a greater sulphate content.

agarose polysaccharide; together with agaropectin makes up agar. Agarose is made up of galactose groups.

alanine amino acid with a sweetish taste.

algae heterogeneous group of organisms that are not necessarily mutually or closely related to plants. Virtually all algae can photosynthesize. Some algae are unicellular microalgae, for example, the bacteria-like blue-green microalgae that, together with diatoms, are classified as phytoplankton (plant plankton). Larger, multicellular algae are called macroalgae or seaweeds. Algae are divided into three groups, somewhat in accordance with their color: green algae, red algae, and brown algae. Classification also covers other characteristics of each of the principal groups. Green algae and red algae are more closely related to each other than to brown algae and green algae are closely related to higher plants, for example, flowering land plants. At the moment, consideration is being given to abandoning the old nomenclature for the plant kingdom, Plantae, where the red and green algae belong, and instead designate this kingdom as Archaeplastida, that is, organisms that have chloroplasts in their original form. They are differentiated among themselves with regard to which type of chloroplast they have. Brown algae, along with different groups of microalgae (including diatoms), are classified as heterokonts, which are distinguished by the special characteristics of their male gametes.

alginate polysaccharide found in brown macroalgae. This substance was discovered in 1886 and was first produced commercially in Norway in 1919 under the trade name Norgine. Alginates are made up of long, linear molecules, which are composed of two different monosaccharide groups, β-D-mannuronate (M) and α-L-guluronate (G). These acid groups can be bound linearly as, for example, -M-M-M-M-M-, -G-G-G-G-G-, or -M-G-M-G-M-G-. The length of the chains varies, with the shortest typically consisting of 500 monosaccharide groups. Alginates form gels in the presence of Ca^{++} (or other divalent ions such as Mg^{++} and Ba^{++}) and gelation takes place at much lower temperatures than it does with pectins. The melting point of alginate gels lies just above the boiling point of water. The gelation process binds very large quantities of water, which is the basis for the use of alginates as thickeners and stabilizers. Alginate gels are also acid resistant, which gives them an advantage over other stabilizers. Because of their water solubility, alginates have many practical applications, especially sodium alginate, which in ionic form is an electrolyte. Calcium alginate, on the other hand, is not water soluble. Alginates are converted to alginic acids by treating them with acids.

alginic acid a mixture of alginates in acid form; insoluble in water.

algotherapy therapeutic treatment using seaweeds and seaweeds products, for example, as body wraps in spas or in cosmetic products.

alkaline having the properties of an ionic salt of, among other substances, an alkali metal such as NaOH or KOH; often used as a synonym for a basic substance. Calcium carbonate ($CaCO_3$) is also said to be alkaline.

alpha-linolenic acid polyunsaturated omega-3 fatty acid with 18 carbon atoms and three double bonds, $(18:3)(9,12,15)$ $CH_3–CH_2–CH=CH–CH_2–CH=CH–CH_2–CH=CH–(CH_2)_7–COOH$. It is the starting point for the formation of superunsaturated fatty acids in the omega-3 family, e.g., DHA (docosahexaenoic acid) and EPA (eicosapentaenoic acid).

amino acid a small molecule made up of between 10 and 40 atoms, which in addition to carbon, hydrogen, and oxygen always contains an amino group $–NH_2$. Amino acids are the fundamental building blocks of proteins. Examples include glycine, glutamic acid, alanine, proline, and arginine. Nature makes use of 20 different, specific amino acids to construct proteins, which are chains of amino acids bound together with so-called peptide bonds. Short chains are called polypeptides and long ones proteins. In food, amino acids are often found bound together in proteins and also as free amino acids that can have an effect on taste. An example is glutamic acid, which is the basis of the basic taste, *umami*. Of the 20 natural amino acids, there are nine, known as the essential amino acids, that cannot be produced by the human body and that we must, therefore, obtain from our food (valine, leucine, lysine, histidine, isoleucine, methionine, phenylalanine, threonine, and tryptophan).

amphiphile a substance or molecule that has mixed feelings toward water. Typically used to describe molecules, such as proteins and fats, that consist of two parts, one of which attracts water, while the other repels it.

amylose polysaccharide consisting of long, linear chains of glucose molecules; with amylopectin, it is the most important ingredient in starch.

amylopectin polysaccharide consisting of a branched network of glucose molecules; with amylase, it is the most important ingredient in starch.

antibacterial related to the ability of a substance to kill bacteria or slow their growth.

antibiotics substances that fight microorganisms such as bacteria and fungi. Penicillin is an antibiotic. Seaweeds and algae contain many substances that have an antibiotic effect, such as iodine, tannic acid, polyphenols, acrylic acid, and bromophenols. These have an antibacterial effect

and help to discourage herbivores from eating the algae.

antioxidant substance that prevents oxidation of other substances. Ascorbic acid (vitamin C), vitamin E, and green chlorophyll are important antioxidants in foodstuffs. DMSP (dimethylsulfoniopropionate), found in green and red algae, is thought to play an important antioxidant role in maintaining the physiological functions of the algae. Carotenoids often function as antioxidants.

anti-viral related to the ability of some substances to suppress the spread of viruses and virus-dependent illnesses. Seaweeds contain substances that work to prevent a virus, such as HIV, from invading the body. Carrageenan, extracted from red algae, seems to be especially effective as an anti-viral agent.

ao-nori flakes of green seaweeds, such as *Monostroma* or *Ulva*; see also *hitoegusa*.

apoptosis programmed cell death.

aquaculture farming of marine organisms, such as fish, shellfish, and seaweeds, in water under controlled conditions. Multi-trophic aquaculture involves several species, e.g., fed fish together with seaweeds that absorb inorganic nutrients and nitrogen from the fish, and/or filter-feeding animals, such as shellfish, that absorb the organic waste from the fish.

arachidonic acid (AA) superunsaturated long-chain fatty acids with 20 carbon atoms and four double bonds, $(20:4)(5,8,11,14)$ $CH_3–CH=CH–CH_2–CH=CH–CH_2–CH=CH–CH_2–CH=CH–(CH_2)_7–COOH$; belongs to the omega-6 family.

Archaeplastida group of photosynthesizing eukaryotes consisting primarily of land plants and red and green algae. It was formerly referred to as the plant kingdom, Plantae, and encompasses all organisms with primary chloroplasts.

arsenic chemical element (As). Found in brown algae and especially abundant in *hijiki*. Arsenic, particularly inorganic arsenic, such as arsenite and arsenate, is poisonous in large doses. Organically bound arsenic, for example, in arsenoribosides and arsenobetaine, is more readily excreted by the body and is, therefore, less poisonous.

arsenobetaine organic form of arsenic which is the main form of arsenic in fish.

arsenoribosides chemical compounds of sugar (ribose) and arsenic; also called arsenosugars.

ATP see adenosine triphosphate.

ayurveda ancient, traditional form of medicine and health care native to India, which incorporates the use of seaweeds; still practiced extensively in India and Nepal.

azuki small red or green beans cultivated in Japan. The red variety has a sweetish taste and is, therefore, used in the form of a paste in Japanese cakes, desserts, and confections.

benthic algae algae that attach themselves to the seabed, stones, or the surface of other organisms, such as bivalves. By far the majority of seaweeds are benthic algae.

beta-carotene the precursor of vitamin A, found in fruits and vegetables, for example, in carrots and in seaweeds, especially *Porphyra* and *Alaria*. The body converts beta-carotene to vitamin A as needed.

bile alkaline fluid mixture produced in the liver and stored in the gall bladder, from which it is secreted into the duodenum where it helps to emulsify the fats in food. The mixture consists of cholesterol derivatives, fats, and a number of salts.

bioavailability the proportion of a drug, supplement, or foodstuff that reaches the bloodstream, whether ingested, administered orally or topically, or injected intravenously.

bioethanol ethanol produced from biological substances, e.g., plants, corn, algae, and seaweeds. The ethanol is produced by fermentation or, alternatively, by enzymatic conversion of the carbohydrates in the biomass.

bromelain enzyme extracted primarily from pineapples; can break down proteins, e.g., those in gelatine, and widely used for tenderizing meat.

bromophenols bromine-containing phenol compounds that accumulate, for example, in saltwater fish and the algae on which the fish or their prey feed. In greatly diluted concentrations, bromophenols are responsible for the smell we associate with a fresh sea breeze.

calcium alginate water insoluble calcium salt derived from alginic acid.

carbohydrates saccharides or sugars that make up a large group of organic compounds that primarily contain oxygen, hydrogen, and carbon. Simple saccharides, that is, monosaccharides and disaccharides, are sweet and encompass the ordinary sugars, such as glucose, fructose, and galactose, as well as saccharose, lactose, and maltose. Starch, cellulose, and glycogen are polysaccharides that are well-known from terrestrial plants. Seaweeds contain polysaccharides such as agar, alginate, and carrageenan. Carbohydrates are produced in plants and algae by photosynthesis, a process in which carbon dioxide and water combine, giving off oxygen.

carbon dioxide gas consisting of CO_2 molecules.

carbon monoxide poisonous gas consisting of CO molecules.

carnivore flesh eater.

carotenoid any of a group of yellow, orange, red, and brownish pigments found in algae, plants, and animals, e.g., fucoxanthin in brown algae and carotene in carrots. Carotenoids can act as antioxidants.

carpospore diploid spore produced in the spore case of red algae, for example, in *Porphyra*, where the carpospores sprout to form the *Conchocelis* stage.

carrageen *Chondrus crispus*. Red alga containing substantial amounts of carrageenan.

carrageenans polysaccharides found in red algae, composed of the monosaccharide galactose in compounds containing varying quantities of sulphate. Carrageenans are long, flexible chains with about 25,000 galactose components. Their gelation properties are dependent on the variety and their behavior is affected by the surrounding conditions of pH, ion content, and temperature. Carrageenans are so-called electrolytes that are able to form gels in the presence of K^+ and Ca^{++} ions. Some can curl into helical structures that are able to link loosely together to form a network. For technical end uses there are three important carrageenans: κ-carrageenan (kappa), which forms strong, stiff gels; ι-carrageenan (iota), which forms softer gels; and λ-carrageenan (lambda), which is well-suited for emulsifying proteins. Only λ-carrageenan is soluble in cold water. Carrageenans function in a bioactive way and are thought to help reduce the risk of cervical cancer by preventing the penetration of viruses.

cell the smallest living entity of an organism. Each cell is protected from its surroundings by a membrane, which forms part of the cell wall. The cells of higher organisms, the eukaryotes, are about 10 μm across and contain a nucleus in which the cell's genetic material is encapsulated. Certain species of seaweeds have very large, long cells that are up to several centimeters in length and contain several nuclei. Some organisms, such as bacteria and yeast, are unicellular, while others are multicellular, consisting of a few hundred to billions of cells. A human is made up of about 100,000 billion cells.

cellulose polysaccharide, made up of linear chains of beta-glucose, as opposed to the chains of alpha-glucose found in starch. Cellulose is found in the cell walls of plants and seaweeds. While the enzymes in human stomachs and intestines are able to break down alpha-glucose, they do not work on beta-glucose, which is why we are unable to digest cellulose.

chelation term used to describe the chemical process by which two or more chemical groups grasp onto, or bind with, each other to form a compound. It is typically used to denote the way in which a metal ion is bound to different ligands, e.g., iron in hemoglobin or magnesium in chlorophyll. In seaweeds, minerals can be chelated to amino acids and colloids, which helps to enhance the bioavailability of the minerals in the intestines.

chirashi-**zushi** sushi that consists of a layer of cooked rice on top of which an assortment of fish, shellfish, omelette, seaweeds, and vegetables pieces are placed.

Chlorella edible microalgal genus ,which includes plant-like, single-celled eucaryotes. High in protein and omega-3 fatty acids.

chlorophyll the pigment in the chloroplasts of plants, seaweeds, and algae that helps to capture light, enabling the organism to carry out photosynthesis. There are several types of chlorophyll. Chlorophyll *a*, which is green, is found in all species of seaweeds and absorbs light from the red and blue portions of the electromagnetic spectrum, with the result that the organism appears green. In addition, the green seaweeds contain chlorophyll *b*, the brown ones, chlorophyll *c*, and the red ones chlorophyll *d*. These other types of chlorophyll absorb light of other wavelengths, changing the appearance of the seaweed. For example, chlorophyll *b* results in a more yellowish color. Chlorophyll is soluble in fats.

Chlorophyta the division of the green algae; together with terrestrial plants makes up the phylum Viridiplantae.

chloroplast organelle in plants and algae containing the green pigment chlorophyll *a* that captures the light energy used for photosynthesis. Chloroplasts in green and red algae are surrounded by a double membrane, whereas those in brown algae are encased by four.

cholesterol fat found in large quantities in the cell membranes of animals. It is the precursor for the formation of sex hormones, vitamin D, and bile salts. Generally, only a little cholesterol is present in seaweeds, which instead avail themselves of other sterols such as desmosterol and fucosterol.

chromosome macromolecular structure that transmits the genetic material of an organism. A human has 23 chromosome pairs that are composed of about 25,000 genes. Chromosomes are long strands of DNA that are coiled around a specific type of proteins.

coagulation process whereby a substance becomes thickened or gathers together, for example, to form a blood clot. Some substances bring about the formation of blood clots, while others, such as plasmin and heparin, tend to inhibit their formation or dissolve them. Fucosterol and fucoidan act as anti-coagulants by increasing the production of plasmin and heparin, respectively.

colloids particles that are so minute that they can remain suspended in a liquid, for example, fat globules in homogenized milk or clay particles in a glacial lake. The particles and the medium in which they are dispersed are referred to as a colloidal system. Many minerals and trace elements found in seaweeds are in colloidal or chelated form.

complex fluid liquid that contains molecular aggregates or long molecules, for example, polysaccharides in ketchup, which impart unusual fluidity properties. An aqueous solution of seaweed polysaccharides, such as agar, alginate, and carrageenan, can produce a complex, viscous fluid. Liquid crystals and polymers often form complex fluids.

Conchocelis the diploid, microscopic stage in the life history of seaweeds such as *Porphyra*, which takes place on, and in, bivalve shells.

conchospore haploid spores that grow, for example, in the *Conchocelis* stage of *Porphyra*. The conchospores are released and deposited onto a suitably stable substrate where they grow into the sexually differentiated

Glossary

261

haploid generation, the large blades of which are harvested to produce *nori*.

Courtois, Bernard French chemist (1777–1838) who, in the course of producing saltpetre from seaweed ash, discovered the element iodine.

Crawford, Michael A. British neurochemist and brain researcher.

Cryptogamia archaic botanical classification of those plants that lack stems and true flowers, for example, algae, mosses, lichens, fungi, and ferns.

Cunnane, Stephen Canadian researcher studying human evolution.

dashi soup stock made from an extract of bonito fish flakes (*katsuobushi*) and *konbu*. *Ichiban dashi* and *niban dashi* refer to the first and second extracts, respectively.

deoxyribonucleic acid (DNA) polynucleotide consisting of a chain of nucleic acids; basis for the genetic information encoded in genetic material and the genome. In the genome, DNA forms a double helix in which two DNA chains spiral around each other.

desmosterol higher-level sterol found especially in red algae; related to cholesterol in animals.

DHA see docosahexaenoic acid.

diatoms unicellular microalgae, encased in a silicate shell with two valves; one of the most common types of phytoplankton (plant plankton).

dietary fiber macromolecules that cannot be broken down by the body's enzymes. The distinction is made between fibers that are water soluble and those that are not. Seaweeds contain both types: for example, the polysaccharides agar, alginate, and carrageenan are soluble, while cellulose and xylan are not. Dietary fiber can constitute up to 75% of the dry components of seaweeds.

diffusion random movement (Brownian motion) of molecules or small particles.

dillisk Gaelic name for dulse (*Palmaria palmata*).

dimethyl sulphide (DMS) organic compound containing sulphur, CH_3SCH_3, formed by oxidation or organic decomposition of dimethylsulfoniopropionate (DMSP) in green and red algae. DMS has a characteristic pungent odor given off, for example, by heated milk, cabbage, cooked mussels, and rotting seaweed. The formation of DMS in the oceans is deemed to be an important

factor in the climate change that is thought to be taking place on Earth. When DMS is released in the atmosphere, it is oxidized to create compounds that, in the form of aerosols, can cause condensation of the atmospheric moisture content, thereby promoting cloud formation and having an effect on the weather.

dimethylsulfoniopropionate (DMSP) organic compound containing sulphur, $(CH_3)_2S^+CH_2CH_2COO^-$, found in red and green (but not brown) algae, where it serves to maintain the proper osmotic balance between the organism and the surrounding salt water. DMSP, which has neither odor nor taste, possibly also acts as an antioxidant within the cells of the seaweeds. When decomposing by oxidation or bacterial action, especially in the presence of the genus *Roseobacter*, DMSP in dead algae and seaweeds gives off the foul smelling gas, dimethyl sulphide (DMS). Fish and shellfish that feed on seaweeds and algae accumulate DMSP in their cells.

diploid word used to describe cells that have the normal complement of chromosomes, that is, a double set.

DMS see dimethyl sulphide.

DMSP see dimethylsulfoniopropionate.

DNA see deoxyribonucleic acid.

docosahexaenoic acid (DHA) superunsaturated, long-chain fatty acid with 22 carbon atoms and six double bonds; member of the omega-3 family. Found in only very small quantities in seaweeds, but in great abundance in microalgae, for example, Spirulina.

domoic acid poisonous amino acid, which acts as a neurotoxin, found in some microalgae. It can accumulate in shellfish that feed on microalgae and is the principal cause of shellfish poisoning.

Drew-Baker, Kathleen Mary British alga researcher (1901–57), whose discoveries regarding the life cycle of the red alga *Porphyra* were the scientific foundations for the large-scale cultivation of *Porphyra* for *nori* production.

Druehl, Louis D. Canadian seaweed researcher.

eicosanoid hormone or signalling molecule formed from omega-3 or omega-6 fatty acids, which play an important role in the regulation of such functions as blood flow and the immune defenses.

eicosapentaenoic acid (EPA) superunsaturated, long-chain fatty acid with 20 carbon atoms and five double bonds; member of the omega-3 family; found in great quantities in seaweeds and algae.

emulsion mixture consisting of a polar liquid, typically water or an aqueous solution, in which an oil-like substance, for example, a fat that is only sparingly soluble in it, is dispersed in small droplets. Emulsification can be enhanced with emulsifiers, which are substances that can bind oil and liquid together, e.g., amphiphiles such as lipids. Emulsifiers lower the surface tension between oil and polar liquids. Mayonnaise and ice cream are examples of emulsions.

enzyme protein that functions as a catalyst for a chemical or a biochemical reaction.

EPA see eicosapentaenoic acid.

epiphyte organism, such as an alga, bacterium, fungus, lichen, or moss, that lives on the surface of a living plant. Examples include the unicellular or small multicellular algae that fasten onto the surface of macroalgae.

estradiol sex hormone belonging to the estrogen group.

estrogens class of sex hormones, of which estradiol is the most important in humans. Estrogens are important sex hormones in women, but are also present in men.

ethanol 'ordinary' alcohol, CH_3-CH_2-OH.

eubacteria group of lower organisms that together with the archaebacteria make up the prokaryotes. Ordinary bacteria are eubacteria.

eukaryote higher organism, either unicellular or multicellular, whose genetic material is enclosed in a nucleus. Fungi, plants, seaweeds, and animals are all eukaryotes. Primitive unicellular organisms that lack a nucleus are called prokaryotes.

fat common designation for an extensive class of substances that are not soluble in water. Fats can be solid, e.g., butter and wax, or liquid, e.g., olive oil and fish oil. The melting point of a fat has a major impact on its taste and nutritional value. A typical fat consists of a long chain of carbon atoms, which can be either saturated or unsaturated. Lipids are an important type of naturally occurring fats. They are composed of fatty acids bound to a variety of other substances such as amino acids and saccharides. Lipids have

mixed feelings toward water as they have two different ends. One is like oil (fatty-acid tail) and hates water, while the other (polar head), which is soluble in water, loves water. Lipids are amphiphilic molecules.

fatty acid compound consisting of a long chain of carbon atoms with a carboxylic acid group. Adjoining atoms in the chain are chemically joined by either a single or a double bond. Those with the most double bonds are described as being the most unsaturated. If only single bonds are present the fatty acid is said to be fully saturated. Monounsaturated fatty acids, e.g., oleic acid from olive oil, have a single double bond. Polyunsatured fatty acids have more than one double bond, such as the two double bonds in linoleic acid from soybeans or three double bonds in alpha-linolenic acid found in flax seed and seaweeds. Superunsaturated fatty acids have more than four double bonds, e.g., the five double bonds in EPA (eicosapentaenoic acid) derived from fish oil and seaweeds. Essential fatty acids are fatty acids that the human body cannot itself produce and that, therefore, have to be obtained from food sources. There are two families of these, both polyunsaturated fatty acids, derived from linoleic acid and alpha-linolenic acid, respectively. They are the progenitors of two important types of fatty acids, the omega-3 and omega-6 fatty acids.

fermentation process in which microorganisms (or microbes), such as yeast or bacteria, convert sugars to an alcohol, e.g., to ethanol, or to an acid, e.g., vinegar.

folate important water soluble vitamin B.

fucoidan a group of sulphated polysaccharides, found especially in *konbu* and *wakame*. Stimulates the formation of heparin, which inhibits the formation of blood clots. Fucoidan may act as an anti-cancer agent in that its sulphur content induces cancer cells to commit suicide in a programmed, controlled manner (apoptosis). It also counteracts the formation of gastric ulcers by suppressing the ability of their precursor bacteria, *Helicobacter pylori*, to colonize the stomach lining. Fucoidan was first isolated from bladder wrack, *Fucus vesiculosus*, hence its name.

fucosterol higher sterol that plays the same role in algae and seaweeds as cholesterol does in animals; found especially in brown algae. The name is derived from the genus *Fucus*.

fucoxanthin carotenoid that is the pigment responsible for the characteristic yellowish brown color of brown algae.

funori red seaweed (*Gloiopeltis* spp.).

furcellaran a type of carrageenan extracted from the red alga, *Furcellaria lumbricalis*. Also called 'Danish agar', although it is not an agar.

furikake condiment often sprinkled on warm rice and other dishes; usually consists of a mixture of salt, dried bits of seaweed (typically *nori*), and fish flakes, as well as toasted black or white sesame seeds.

futomaki thick *maki*-zushi rolls made using a whole sheet of *nori*.

galactose a sugar sweeter than glucose.

gamete reproductive cell, either an ovum or a sperm, containing one copy of the chromosome set. A seaweed organism consisting of cells with a single chromosome set is called a gametophyte.

gamma-linolenic acid polyunsaturated omega-6 fatty acid with 18 carbon atoms and three double bonds, $(18{:}3)(6,9,12)$ $CH_3-(CH_2)_3-CH{=}CH-CH_2-CH{=}CH-CH_2-CH{=}CH-(CH_2)_4-COOH$; starting point for the formation of superunsaturated fatty acids from the omega-6 family, especially AA (arachidonic acid).

gel technical term for a network of molecules that incorporates a great deal of water but is also somewhat stiff like a solid. Seaweeds contain large quantities of polysaccharides, such as agar and carrageenan in the red species and alginates in the brown ones. These are good gelling agents that can form gels that are very strong and more stable than those made with gelatine.

gelatine the same protein as the one found in the form of collagen in animal connective tissue. In contrast to collagen, gelatine is soluble in water and is formed when collagen is heated, dissolving the stiff fibers in it. On cooling, the stiff fiber structure of collagen is not formed again; in its stead a gel containing water, which we know as a jelly, is produced.

gelation process of forming a gel, for example, when egg whites are heated or gelatine is cooled.

gene a sequence of nucleotides of DNA that, among its other functions, contains the genetic information of an organism (hereditary material). The gene complement of an organism is referred to as its genome.

gim Korean word for red algae from the *Porphyra* genus; also known as *kim*.

glucose sugar or monosaccharide, $C_6H_{12}O_6$, that is the most important carbohydrate in plants and animals. In plants and algae, glucose is formed by photosynthesis.

glutamic acid amino acid found in such foods as fish, shellfish, and seaweeds, often in the form of a salt, monosodium glutamate (MSG), which is the basis for *umami* taste.

gluten certain proteins (especially gliadin and glutenin) found in wheat, which enhance the baking properties of dough made with wheat flour. Kneading stretches the proteins and forms an elastic, water-binding network that is ideal for trapping the bubbles of carbon dioxide that are formed when the dough rises.

glycolipid lipid that has carbohydrates bound to the water soluble end (the polar head).

goitre medical condition of enlargement of the thyroid gland; usually attributable to iodine deficiency. Iodine found in foods such as seaweeds, fish, and shellfish are an important factor in the low occurrence of goitre in populations that regularly eat food from marine sources.

guanosine monophosphate (GMP) nucleotide formed together with inosine monophosphate (IMP) when the energy-storing biomolecule ATP is broken down by the cells to produce energy. This substance has *umami* taste and is 10 to 20 times more potent than MSG.

guluronic acid organic acid involved in the formation of alginates.

gunkan-zushi sushi made by enclosing ingredients that might otherwise fall apart in a piece of *nori*; also known as battleship sushi.

Haeckel, Ernst H. P. A. German biologist and philosopher (1834–1919).

haploid word used to describe cells that have only a single set of chromosomes, for example, reproductive cells.

HDL high-density lipoprotein.

heavy metal metallic element with a large mass. Some heavy metals, for example, mercury, cadmium, and lead, are poisonous. Others, for example, iron, molybdenum, manganese, cobalt, and copper, play an important nutritional role as trace elements.

heparin carbohydrate of the glucosaminoglycan type; used as a very effective anticoagulant to prevent blood clots.

herbivore plant eater.

heterokonts phylum of algae encompassing the brown algae and various groups of microalgae (including diatoms) that are distinguished by some special properties of their male reproductive cells. The paradoxical result is that these small, unicellular microalgae have been placed in the same group as the largest macroalgae in the world.

hijiki the Japanese name for the brown macroalga *Sargassum fusiforme*, which is greenish-brown and bushy with long, narrow, elongated blades.

Hirano, Bujiro Japanese who, in 1878, introduced a type of transplantation technique for the cultivation of *Porphyra* in Japan. The technique consists of moving poles with the seaweed sprouts from one area of cultivation to another. The method is purely empirical but makes sense in light of the later discovery that the life cycle of *Porphyra* has two stages that place different requirements on their surroundings.

hitoegusa green algae (*Monostroma nitidum, Monostroma latissimum, Ulva* spp.), also known as *ao-nori*, which are used in Japan in the form of small, dried flakes that are sprinkled on warm rice.

HIV human immune deficiency virus that can cause AIDS.

hominid member of the large ape family that includes humans.

hormone chemical messenger molecule that works in the body over great distances, for example, by being transported in the blood stream from the gland where it is secreted. Examples include two important iodine-containing hormones (thyroxine and triiodothyronine) produced in the thyroid gland, the sex hormone estrogen, and insulin, which regulates sugar metabolism.

hoshi-nori nori produced from the red alga *Porphyra* by air-drying; in contrast to *yaki-nori*, it is not toasted.

hosomaki thin *maki*-zushi rolls made with a half sheet of *nori*.

hydrocarbons organic compounds made up exclusively of carbon and hydrogen, for example, in the form of a chain of carbon atoms (hydrocarbon chain) in oils or fats.

hydrogel a somewhat solid material (gel) which contains a great deal of water. Hydrogels have good properties with respect to stability and viscosity, which can be exploited industrially, for example, to stabilize foods that are liquid. Hydrogels based on seaweed extracts are used widely in the food sector, where they are incorporated into meat, fish, dairy products, and baked goods. Alginates from brown algae, as well as agar and carrageenan from red algae, are well suited for making hydrogels.

immune system the body's defense system that wards off foreign substances and invading organisms, such as viruses and bacteria; there are both innate and acquired immune defenses.

inosine monophosphate (IMP) nucleotide, formed together with guanosine monophosphate (GMP) when the energy storing biomolecule ATP is broken down by the cells to produce energy; has *umami* taste and is 10 to 20 times more potent than MSG.

insulin hormone that regulates the sugar metabolism in the blood stream.

inulin carbohydrate that some plants, such as Jerusalem artichokes, onions, and garlic, use *in lieu* of starch as an energy depot. Inulin cannot be broken down by human digestive enzymes.

iodine element (I), found in large quantities in brown algae, that was discovered in connection with the use of seaweed ash for the production of gunpowder. When bound in organic form, iodine is called organic iodine. Two important hormones containing iodine, thyroxine and triiodothyronine, are produced in the thyroid gland. Common ions of iodine are iodide (I^-) and iodate (IO_3^-). The radioactive isotope of iodine, ^{131}I, is a toxic contaminant that can be accidentally discharged from nuclear power plants.

kahalalide F polypeptide found in seaweeds, where it acts as a chemical defense against herbivorous fish; has anti-viral properties and may be the reason why HIV/AIDS is less prevalent in populations where seaweeds are a major component of the diet.

kainic acid organic acid present in some types of red algae such as dulse; acts as a neurotoxin and, in large doses, can induce muscle cramps.

kaiso Japanese term for edible seaweeds (*kai* means sea and *so* can refer to plant).

kanten see agar.

karengo Maori word for red algae from the *Porphyra* genus.

katsuobushi cooked, salted, dried, smoked, and fermented *katsuo* (a fish related to tuna and mackerel), which is shaved into paper thin flakes; used to make such things as soup stock, *dashi*.

kelp umbrella term for the large brown algae, especially those with large leaf-like blades. The word is derived from the French *culpe*, which in the Middle Ages was more commonly used to refer to seaweed ash than to the seaweeds themselves.

komochi wakame herring roe on kelp.

konbu Japanese name for different types of kelp that are well suited for eating. They belong to the order known as Laminariales, which includes many edible genera. The name Laminariales alludes to the long, thin lamella-like blades that are characteristic of kelp. *Ma-konbu, rausu-konbu, rishiri-konbu,* and *hidaka-konbu* are different qualities of *konbu*. *Oboro (tororo) konbu* are opaque shavings of *konbu* marinated in rice vinegar that are cut from the seaweed blades with a sharp knife; used as an edible packing for rice and other ingredients.

kuragakoi ageing of seaweeds (*konbu*) in cellars.

laminarin polysaccharide of glucose found in brown algae; named after the brown alga order Laminariales.

laverbread spinach-like purée made from red algae from the *Porphyra* genus.

LDL low density lipoprotein.

lichen a small ecosystem made up of microalgae in a symbiotic association with a fungus. In this way both organisms can survive in an environment where neither could exist on its own.

limu Polynesian expression for seaweeds.

linoleic acid polyunsaturated omega-6 fatty acid with 18 carbon atoms and two double bonds, $(18:2)(9,12)$ $CH_3-(CH_2)_4-CH=CH-CH_2-CH=CH-(CH_2)_7-COOH$; basis for the formation of superunsaturated fatty acids of the omega-6 family, e.g., arachidonic acid.

linolenic acid see alpha-linolenic acid.

lipid amphiphilic fat that consists of a water soluble part and an oil soluble part. Glycolipids are lipids that have a sugar group attached to the polar head.

lipoprotein complex of fats (lipids) and proteins. Lipoproteins, e.g., LDL and HDL, are important for the transport of fats, such as cholesterol, in the body via the bloodstream.

lymphocyte a type of white blood cell, for example, a T-cell, that is a component of the body's immune system.

maccha powdered green tea.

macroalgae a diverse group of multicellular organisms, which can form enormous 'forests' in the ocean. The marine macroalgae are referred to as seaweeds. They are traditionally classified into three groups: brown algae (Phaeophyceae), green algae (Chlorophyta), and red algae (Rhodophyta). As all of the groups contain chlorophyll granules, their characteristic colors are derived from other pigments. Although some of the larger ones have complex structures with special tissues that provide support or transport nutrients and the products of photosynthesis, others are made up of cells that are virtually identical. The smallest macroalgae are only a few millimeters or centimeters in size, while the largest routinely grow to a length of 30 to 50 meters.

macrobiotics a lifestyle concept that encompasses the pursuit of a well-balanced, natural diet that benefits the body in terms of overall wellness and quality of life. The diet is rich in cereals and vegetables, which are processed as little as possible, and can be supplemented with fish, nuts, and fruit.

maërl umbrella term for occurences of loosely-lying calcified cell walls of dead coralline red algae. In the course of the centuries, the dead algae are deposited in a coral-like fashion. The surfaces of these calcareous deposits have been scraped off for use as fertilizer.

maki-zushi sushi roll with a sheet of *nori* either on the outside or the inside; *maki* means to roll in Japanese.

mannitol sugar alcohol found, for example, in seaweeds, especially sugar kelp, and fungi, such as mushrooms. It has half as many calories as sugar, is half as sweet, and is a chemical isomer of sorbitol, which is present in red algae. Mannitol imparts its characteristic sweet taste to seaweeds and helps them to maintain the correct osmotic balance in salt water.

mannuronic acid organic acid involved in the formation of alginates.

mekabu wakame sporophylls.

membrane the boundary between a cell and its surroundings. This term is used particularly to refer to the double layer of lipids (fats) that forms the middle part of the cell wall.

metabolic syndrome composite of lifestyle dependent, non-communicable diseases that are attributable to diet, especially cardiovascular disorders, obesity, type 2 diabetes, high blood pressure, and possibly psychiatric disorders.

metabolite substance that is an intermediate product of an organism's metabolism. Primary metabolites are substances that have an impact on the organism's growth and reproduction, for example, iodine, minerals, and trace elements. Secondary metabolites have important ecological functions, acting, for example, as pigments or bioactive substances.

methane simple organic gas, CH_4; can be formed when anaerobic bacteria break down organic material from plants, animals, and algae.

methyl mercaptan sulphur-containing organic compound, CH_3SH; noticeable as a disagreeable odor, which resembles the smell of rotten cabbage, associated with the decomposition of brown algae and certain types of red algae.

microalgae unicellular algae, for example, the blue-green microalgae, which resemble bacteria; also known as phytoplankton or diatoms.

mirin sweet rice wine with ca. 14% alcohol; used in Japanese food preparation, but not intended to be drunk.

miso **soup** soup made with *dashi*, a stock based on *konbu* and dried fish flakes, to which *miso*, a paste made from fermented soybeans or a cereal grain, such as rice or barley, is added.

mochi soft Japanese rice cake made from steamed white rice that is pounded together to form an elastic paste; see also *senbai*.

monosaccharides simple sugars (carbohydrates), such as glucose, fructose, and galactose, that are made up of a single molecule; usually have a sweet taste.

monosodium glutamate (MSG) sodium salt of the amino acid glutamic acid, also known as 'the third spice' because it is the one most widely used after salt and pepper; imparts *umami* taste.

mozuku brown seaweed (*Cladosiphon oka-muranus*) grown around the islands of Okinawa in Japan.

MSG see monosodium glutamate.

nattō whole, small soybeans fermented to form a stringy, viscous mass with a distinctive flavor, a strong aroma, and an intense *umami* taste. Traditionally eaten in Japan for breakfast with warm cooked rice and raw eggs; can also be added to soups.

niacin vitamin B_3.

nigiri-**zushi** hand shaped sushi made of small rice balls topped with such items as slices of raw fish, shellfish, or pieces of omelette. *Nigiri* means to grasp or hold tightly with the hand.

nitrate chemical compound that contains a nitrate ion, NO_3^-. Nitrates are nitric acid salts.

nitrogen element (N) which is found in all amino acids and, hence, in proteins.

nori paper thin sheets made from the blades of the red alga *Porphyra*, which are dried (*hoshi-maki*) or sometimes toasted (*yaki-nori*). Among other uses, they are essential for making sushi rolls.

nucleic acid chemical designation for a macromolecule made up of a nucleobase (adenine, guanine, cytosine, uracil, or thymine), a monosaccharide, and a phosphoric acid. They are the building blocks in DNA, RNA, and genomes.

nucleotide chemical group that is a component of nucleic acid. The *umami* taste substances guanosine monophosphate and inosine monophospate are nucleotides. DNA is a polynucleotide.

nutraceutical an ingestant (a substance taken into the body by mouth or through the digestive system) or a substance extracted from the ingestant, that has a medicinal or health-promoting effect, either by preventing or treating a disease. In some contexts, algae and seaweed extracts are designated as nutraceuticals.

ogonori red seaweed, sea moss (*Gracilaria* spp.).

Ohmiya, Jinbei Japanese who, in 1821, invented the 'pole system' of cultivating *Porphyra* for *nori* production. It is a simple method that utilized split bamboo stems or bamboo branches. The poles were stuck into sandbeds in shallow water so that they were partially above the surface at low tide. This ensured that the *Porphyra* that fastened onto the poles and twigs had better growing conditions.

oil chemical compound containing carbon that is not soluble in water; examples include fatty acids and lipids.

oleic acid monounsaturated fatty acid with 18 carbon atoms, $CH_3-(CH_2)_7-CH=CH-(CH_2)_7-COOH$; main component of olive oil.

omega-3 fats polyunsaturated fats derived from alpha-linolenic acid, e.g., DHA (docosahexaenoic acid) and EPA (eicosapentaenoic acid).

omega-6 fats polyunsaturated fats derived from linoleic acid, e.g., AA (arachidonic acid).

onigiri a ball of sushi rice wrapped in *nori*.

osmosis process of diffusion of water across a barrier, for example, a cell membrane, that is permeable to water but impermeable to other larger molecules, such as salts, amino acids, or sugars. An imbalance is created when some of the water passes across to the side of the membrane containing the large molecules. The extent of water permeation increases with the degree of hydrophilicity of these molecules. The osmotic effect is counterbalanced by a pressure, called the osmotic pressure, across the entire membrane. Osmosis is central to the ability of plants to draw water from the ground, into their root system, and up through their trunks and branches. The opposite process, known as reverse osmosis, in which pure water is drawn out of a solution, is used for purifying water, e.g., desalination.

oxidation removing one or more electrons from an atom, ion, or molecule. For example, the double bonds of unsaturated fats can be oxidized, resulting in rancidity.

papain enzyme found in papaya that is able to breakdown proteins, for example, those in gelatine. As connective tissue in meat is made up of gelatine, papain is used as a tenderizer.

pathogen disease-causing organism, for example, a bacterium or a fungus.

pectin polysaccharide derived from plants; used as a thickener in jams and jellies.

Phaeophyceae the class of brown algae. Before it became obvious that the brown algae are a class within the phylum Heterokontophyta, they were classified as Phaeophyta.

phenols large group of acidic chemical substances derived from phenol (hydrobenzene); found in some plants and seaweeds, to which they impart their characteristic slightly bitter taste with hints of grass, hay, flowers, and smoke. Bromophenols accumulate in fish that consume marine algae

directly or prey on smaller fish that do. The smell of bromophenols is associated with that of a fresh ocean breeze.

pheromone substance that emits chemical signals that an organism can use to announce its presence to others of the same species, for example, to attract a partner for reproduction.

phosphate chemical compound containing one phosphate ion, PO_4^{3-}. Phosphates are phosphoric acid salts.

photosynthesis process whereby plants and algae, using energy from the sun, convert carbon dioxide and water into carbohydrates, such as glucose, and at the same time release oxygen as a waste product. Photosynthesis takes place in the chloroplasts, which contain chlorophyll that absorbs the sunlight. All algae and seaweeds carry out photosynthesis and are, thereby, responsible for producing a significant proportion of the oxygen found in the Earth's atmosphere.

phycobilins type of water soluble pigments that are the reason for the red, orange, and blue appearance of red algae and cyanobacteria.

phycology the study algae, including seaweeds; derived from the Greek word *phycos*, which means seaweed.

phytoplankton unicellular microalgae; also called plant plankton.

pigment a coloring substance, for example, the green chlorophyll *a* in seaweeds and plants. Different algae have pigments that are suited to capturing the light of varying wavelengths that is available in their surroundings and that result in their brown, yellowish, or red appearance. In the brown algae, the green color of the chloroplasts is covered by a brown pigment, fucoxanthin, which is a carotenoid.

plant plankton see phytoplankton.

Plantae the plant kingdom (Archaeplastida).

plasmin blood enzyme that helps to prevent the formation of blood clots by dissolving the fiber protein fibrin.

polyelectrolyte ionic molecule with many charges. Sodium alginate from brown algae is an example of a polyelectrolyte whose electric properties are central to its abilities for forming gels.

polymer large molecule, either in the form of a chain or branched structure, composed of many identical or different units (monomers). An example is a protein, in the form of a polyamide, made up of amino acids, or a polysaccharide, composed of many sugar groups. Polymers can be made by a polymerization process in which the individual monomers are bound together in a chemical reaction.

polypeptide chain of amino acids bound together by peptide bonds. Short chains are called polypeptides and long ones proteins.

polyphenol chemical compound containing several phenol groups; often components of the bitter substances in brown algae. See also tannin.

polysaccharide see also carbohydrate; sugar composed of several saccharide units. Up to 40% of the dry weight of seaweeds can consist of polysaccharides, including the group known as sugar alcohols. Polysaccharides function as the energy store within the cells and as the building blocks of the cell walls themselves as well as of the larger structures such as stipes and blades. The polysaccharides used as structural elements by seaweeds are much more complex and heterogeneous than the simple polysaccharides, such as glycogen and starch, used by plants to store energy. While plants avail themselves of a type of polysaccharides called pectins, which are known to us as thickeners for jams and jellies, seaweeds use polysaccharides that are unique to them, namely, alginate, carrageenan, and agar, which are also called soluble dietary fiber. These fibers can absorb water in the stomach and intestines and form gels. The insoluble dietary fiber in seaweeds is also composed of polysaccharides, such as cellulose and xylan in red and green algae and cellulose in the brown ones. Typically, the insoluble dietary fiber makes up only 2–8% of the seaweeds. Another seaweed polysaccharide is fucoidan, typically found in *konbu* and *wakame*.

potash potassium carbonate, K_2CO_3.

potassium carbonate K_2CO_3, potash.

potassium chloride KCl.

potassium nitrate KNO_3, saltpetre.

prokaryote unicellular organism that lacks a nucleus. All bacteria, as well as the blue-green algae, are prokaryotes.

protein polyamide, that is, a long chain of amino acids bound together by peptide bonds. Proteins lose their functional ability (denature) and their physical properties change when they are heated or exposed to salt or acid (for example, when cooked, salted, or marinated). Enzymes are a particular class of proteins, whose function is to ensure that chemical reactions take place under controlled circumstances.

protist older classification used for a heterogeneous group of organisms that do not fit into the three major eukaryotic kingdoms: animals, plants, and fungi. Whether the protists are unicellular or multicellular has no impact on their classification. Brown algae belong to this group.

Rhodophyta the division of red algae.

saccharide sugar; see carbohydrate.

sashimi sliced raw fish or shellfish.

saltpetre potassium nitrate KNO_3.

seaweeds umbrella term for all types of marine macroalgae, whether of the brown, green, or red varieties. In the vernacular, the term is also used to describe plants that grow partially submerged in the sea, such as eelgrass.

senbai traditional salty-sweet Japanese rice cakes, eaten as a snack. They are made from *mochi* that is baked or toasted over a charcoal fire. *Senbai* made with soy sauce and *mirin* have a slightly spicy taste and often have a small piece of *nori* stuck on, or wound around, them.

Shinto system of beliefs and religious practices indigenous to Japan, rooted in the connection between humans and nature.

shiso leaf mint (*Perilla frutecens*), found in red, green, and green-red varieties.

silicon dioxide SiO_2, quartz.

sloke term for the blades of certain commonly eaten seaweeds, used primarily for those also referred to as laver.

soda sodium carbonate, Na_2CO_3.

sodium alginate water soluble sodium salt of alginic acid.

sodium carbonate Na_2CO_3, commonly known as washing soda or soda ash.

sodium chloride NaCl, table salt.

sodium nitrate $NaNO_3$, Chile saltpetre or Peru saltpetre.

sorbitol sugar alcohol, found, for example, in red algae to which it imparts a characteristic sweet taste. Sorbitol helps the cells of the seaweed to maintain a correct osmotic balance in salt water. Sorbitol has two-thirds as many calories as sugar and is 60% as sweet. Chemically, it is an isomer of mannitol, which is found in brown algae.

soy sauce seasoning liquid made from cooked

soy beans that are fermented in a saline solution; called *shōyu* in Japanese.

Spirulina edible blue-green microalgae (*Arthrospira maxima* and *Arthrospira platensis*) that often contain more than 50% proteins and 25% convertible carbohydrates, as well as the same minerals and vitamins as are found in seaweeds. While there is no iodine in Spirulina, it is very rich in iron. It contains about an equal amount of potassium salt and sodium salt, as well as significant quantities of the essential omega-3 and omega-6 fats.

spore cell that functions as a means of dispersal and asexual reproduction for an organism. In contrast to seeds, spores contain very little stored food. By undergoing mitosis, a spore can give rise to a new organism, a multicellular gametophyte, which is still haploid. The gametophyte can produce differentiated gametes (egg and sperm cells) that can fuse and form a zygote, from which a new sporophyte originates. Some seaweeds reproduce by means of spores.

sporophyll the part of the seaweed, or other organism, that produces spores for reproduction.

starch mixture of the polysaccharides amylose and amylopectin.

sterol cyclic carbon compound that consists of a hydrophobic core of four, fused rings. The so-called higher sterols are important for all advanced forms of life (cholesterol in animals, ergosterol in fungi and yeasts, and phytosterol in plants and algae). For example, seaweeds contain fucosterol and desmosterol.

strontium metallic element (Sr). Its radioactive isotope is ^{90}Sr.

sugar see carbohydrate.

sugar alcohol substance resulting from the reduction of the carbonyl group in a sugar. Mannitol and sorbitol are sugar alcohols, found in brown and red algae, respectively.

suimono clear soup made from the first *dashi* (*ichiban dashi*), a soup stock made with *konbu*.

sulphate substance that contains a sulphate group, $-SO_4^{2-}$.

sushi Japanese dish consisting of cooked, vinegared rice, e.g., formed into a ball and covered with pieces of raw fish, shellfish, vegetables, omelette, or fungi.

table salt sodium chloride, NaCl.

tamago Japanese word for egg, usually a chicken egg, but can also be a quail egg. *Tamago-yaki* is a rolled omelette prepared in a special rectangular pan.

tannin (tannic acid) umbrella term for a variety of polyphenolic compounds, which are bitter taste substances, found in red wine, black tea, and smoked products, among others. Brown algae contain certain tannins which make them very bitter if they are overcooked.

taurine amino acid found in significant quantities in some seafoods and in red algae. Major constituent of bile where it functions as an emulsifier to bind fats and mediate the uptake of lipids, e.g., cholesterol. Strictly speaking, taurine is not a real amino acid since it has a sulfonic acid group ($-SO_3H$) instead of a carboxyl group ($-COOH$). The sulphur content of taurine contributes to the characteristic meaty taste of seaweeds.

tazuna-zushi multi-colored (usually red, green, and white) inside-out *maki* roll, where fish and vegetables (either cucumber or avocado) create a special rainbow effect on the outside.

temaki-zushi handrolled sushi, for example, in a cone shape.

tengusa red alga (*Gelidium* spp.) containing agar.

terpenes large group of organic compounds derived from isoprene.

thalassotherapy therapy based on the external application of seawater, seaweeds, and seaweed extracts to the body.

thyroxine important hormone containing iodine secreted in the thyroid gland; together with triiodothyronine controls metabolic functions and growth in the body.

tosaka-nori red seaweed (*Meristotheca papulosa*) that comes in three different colors, white, green, and red.

toxin poison, typically derived from a plant, fungus, or animal. Some blue-green microalgae contain toxins that attack the liver and the nervous system or lead to skin irritations.

trace elements chemical elements, for example, manganese, chromium, selenium, cobalt, zinc, and copper; even though they are consumed in minute quantities, they have great nutritional significance. Seaweeds and algae are sources of the most important trace elements (micronutrients) required by humans.

transgenic term used to describe an organism

that contains a gene or genetic material that have been altered.

triglyceride substance with three fatty acid groups.

triiodothyronine important hormone containing iodine secreted in the thyroid gland; together with thyroxine controls metabolic functions and growth in the body.

trimethylamine foul smelling organic substance (tertiary amine) produced, for example, by bacterial decomposition of trimethylaminoxide in dead seaweeds and fish. Trimethylaminoxide is, by itself, odorless.

tryptophan essential amino acid.

turbulence chaotic movements in gases or liquids, for example, in connection with currents in water.

umami 'the fifth taste' or 'meat taste', brought out especially by MSG; found, for example, in large quantities in *konbu*.

umeboshi dried and brine-pickled Japanese apricots (*ume*) or plums.

uramaki inside-out *maki*-roll that has the sheet of *nori* on the inside and the rice on the outside.

virus a microscopic agent that is smaller than a bacterium; contains its own DNA, which it can use to infect living cells and organisms.

viscosity resistance to flow in a liquid; alternatively, the capacity of a liquid to resist when another substance is moving through it.

vitamin one of a group of different, essential substances that the body itself can produce only in very limited quantities and that are, therefore, primarily derived from the diet. Examples are vitamins A, B, C, D, E, and K. Vitamin C (ascorbic acid) and vitamin E are also important antioxidants. Seaweeds are rich in vitamins, especially A, B, C, and E.

wakame Japanese name for brown macroalgae from the genus *Undaria*, which belongs to the order Laminariales. *Wakame* is, therefore, also a variety of kelp. It is best known from its use in the preparation of almost every type of *miso* soup. *Wakame* has a beautiful dark green color and a rather mild taste. *Hiyashi-wakame* is *wakame* salad prepared with, e.g., sesame and chili.

wasabi Japanese horseradish (*Wasabia japonica*).

xylan polysaccharide that functions as a structural element of the cell walls in red and green algae. It is an insoluble dietary fiber.

yaki-nori toasted *nori* sheets.

yōkan Japanese confectionery or candy based on red *azuki* bean paste sweetened with sugar and thickened with agar (*kanten*) to form a solid jelly.

zicai Chinese expression for *Porphyra* and *nori*.

◄ Dried specimen of the red alga *Scinaia furcelleta* from the collections of the Natural History Museum in London.

Index

Index

◄ Roll of *konbu* for
preparation of soups.

OLE G. MOURITSEN is a scientist and professor of biophysics at the University of Southern Denmark where he is director of MEMPHYS-Center for Biomembrane Physics. He is an elected fellow of the Royal Danish Academy of Sciences and Letters, the Danish Academy of Technical Sciences, and the Danish Gastronomical Academy. His research is directed towards a broad range of basic science questions as well as their applications within biotechnology, biomedicine, and food science. He is the recipient of a number of prestigious prizes for his work, most recently the Danish National Prize for Research Communication (2007), the British Royal Society of Chemistry Bourke Award (2008), and the European Lipid Science Award (2011).

In his spare time, the author cooks and furthers his knowledge of food science and gastronomy. In addition he writes popular articles and books about the science of cooking and taste, often in collaboration with well-known chefs.

In addition to numerous scientific articles, Ole G. Mouritsen is the author of *Life-as a Matter of Fat* (2005), *Sushi. Food for the Eye, the Body & the Soul* (2009), and *Umami* (2011).